Shared Governance for Sustainable Working Landscapes

Shared Governance for Sustainable Working Landscapes

Timothy M. Gieseke

CRC Press
Taylor & Francis Group
Boca Raton   London   New York

CRC Press is an imprint of the
Taylor & Francis Group, an **informa** business

CRC Press
Taylor & Francis Group
6000 Broken Sound Parkway NW, Suite 300
Boca Raton, FL 33487-2742

First issued in paperback 2017

ISBN 13: 978-1-138-49552-4 (pbk)
ISBN 13: 978-1-4987-1800-4 (hbk)

Library of Congress Cataloging-in-Publication Data

Names: Gieseke, Timothy M.
Title: Shared governance for sustainable working landscapes / Timothy M. Gieseke.
Description: Boca Raton : CRC Press, 2017. | Includes bibliographical references and index.
Identifiers: LCCN 2016016909 | ISBN 9781498718004 (hardcover : alk. paper)
Subjects: LCSH: Sustainable agriculture. | Land use. | Land use--Economic aspects. | Agricultural education.
Classification: LCC S494.5.S86 G54 2017 | DDC 338.1--dc23
LC record available at https://lccn.loc.gov/2016016909

Visit the Taylor & Francis Web site at
http://www.taylorandfrancis.com

and the CRC Press Web site at
http://www.crcpress.com

Contents

Section III: Designing a glocal business ecosystem

Preface

In 2004, after a decade of a consistent mix of policy, research, and practitioner experience in the emerging field of agriculture sustainability, I embraced the idea that an *environmental market signal* would be needed to resolve this complex issue. The basis for supporting such a signal is that farmers and others working directly with the land to produce food and fiber commodities are often delegated to the role of price-taker and policy-taker. It is the efficacy of a market signal that is able to speak to the countless variations of production systems within the dynamics of economic, climatic, and other uncontrollable forces that dominant the land practitioner's livelihood.

From 2005 onward I was able to apply the components of this *market signal* concept to several projects at the local, regional, national, and international levels. This compelled me to write of its effects and consequences in a 2011 book, *EcoCommerce 101: Adding an Ecological Dimension to the Economy*. This eco-commerce model garnered interest from many sectors at all levels, but it failed to gain traction or offer a path toward enabling transactions associated with such an environmental market signal.

Shortly after *EcoCommerce 101* was published, I was intrigued by a shared governance model I applied to a state-level environmental quality assurance program for livestock farms. A spreadsheet acting as a crude platform revealed how disparate sustainability stakeholders could symbiotically achieve their objectives at lower cost than if they independently pursued their goals.

As technology improved and people became more interconnected, multisided platforms became a business reality and supported the evolution of e-commerce ecosystems. The next step to take, as I began writing this manuscript, was to build a multisided shared governance platform that could incorporate an environmental market signal and support transactions. With this platform model, I was able to envision the potential and emergence of an eco-commerce ecosystem.

Acknowledgments

I would like to acknowledge the many insights provided by professionals, practitioners, and policy-makers seeking sustainable solutions from the landscape over the many decades. It is these countless efforts that shape technology, science, policies, and new ways of thinking. The wisdom in Elinor Ostrom's statement, "If it works in practice, it must work in theory" provides the observer and conscientious practitioner with the confidence to continue forward and begin to understand society's *natural* solutions.

I would like thank my wife, Jenny, and our three boys, Max, Isaac, and Eli, for their support and patience. I would also like to thank friend and graphic designer Ron Schrader for the artwork.

I would also like to acknowledge the interdisciplinary contributions of several colleagues and the moral support they provided as I continued on this path: Melinda Kimble, United Nations Foundation; A.G. Kawamura, Southern California farmer, former California Secretary of Agriculture; Louis Meuleman, European Commission; Dr. Cornelia Flora, Iowa State University; Gabriel Thoumi, University of Maryland; Jerry Hatfield, USDA ARS; Dan Abelow, Expandiverse; and Ira Feldman, University of Pennsylvania.

Exploring new territory is of course challenging, and the first solutions applied are seemingly fragile. Their thoughtful, inquiring, and sincere words were most appreciated.

Author

 **Timothy M. Gieseke**'s interdisciplinary career is reflected in the research and insights of his writings. A master's degree in environmental sciences is a cornerstone for his perspective on agriculture sustainability. He also brings experience in agriculture production, governmental experience in conservation planning, policy analysis at state and federal levels, political endeavors, and agribusiness management.

With this near panoramic view of landscape sustainability, Tim recognized the need for a transdisciplinary approach to enable practitioners and policy-makers to transcend and blur the lines between their traditional organizational boundaries. He has carried this vision through several of his local to global efforts.

chapter one

Introduction

The world would be a different place without the benefits received from agriculture and food systems (Müller et al., 2015). These agroecosystems, which are a biological and natural resource system managed by humans for the primary purpose of producing food, could also be considered the largest ecological experiment on Earth (Sandhu et al., 2015). For millennia, agriculture has been the most visible example of human interaction with the landscape (Wood et al., 2000).

Some 26% of the earth constitutes ice-free land. Of that, land use consists of urban (1%), villages (6%), croplands (21%), rangelands (30%), forests (19%), and wildlands (23%). These land uses are in transition from wildlands, initial clearing, subsistence, and small-scale farming, to urban settlement and intensive agriculture; all of these stages are going on concurrently somewhere in the world (Defries et al., 2006).

Agroecosystems now cover 5 billion hectares worldwide and include 1.5 billion hectares arable land under annual crops, such as cereals, legumes, and oilseed crops (Wratten et al., 2013). At the turn of the century, the food production from these agroecosystems was valued at around $1.3 trillion per year and provided 94% of the protein and 99% of the calories consumed by humans. This production system directly employs around 1.3 billion people (Wood et al., 2000).

1.1 Agricultural transition phases

Agricultural systems have transitioned through four phases in the last two centuries from primarily a labor-intensive industry that converted natural landscapes to farmland to a high-tech production system influenced by a socioeconomic movement seeking environmental benefits (Timmer, 1988). Each subsequent phase of agriculture production was accompanied by an increase in the number and diversity of stakeholders resulting in a more complex socioeconomic system.

The pioneering phase began during the late nineteenth century in North America, Australia, and New Zealand. This first phase was characterized by critical shortages of labor and capital and involved land clearing and development. A relatively large percentage of the population was involved directly by providing labor through human capital. Stakeholders consisted of farmers, governments, industry, and localized trade.

Second, the production phase occurred in the early portion of the twentieth century. The growth of both domestic and international markets led to the use of technologies and science to improve farming systems. Additional stakeholders became involved as industries provided more production inputs, governments provided support, land grant universities expanded research, and businesses expanded trade.

Third, the productivity phase in the mid- to late-twentieth century focused on improving the productivity of farming systems to address the long-term decline in real commodity prices. Industrialization reduced the need for human labor and the sphere of stakeholders grew because inputs of energy, fertilizers, genetics, chemistries, and processing techniques were researched and developed. The conservation movement emerged with a focus on soil erosion and addressing on-farm conservation needs. This phase contained similar stakeholder types of the second phase but the numbers increased owing to its industrialization in production and conservation. This industrialization of agriculture led to significant positive economic effects, but negative ecological effects were beginning to be realized (Sandhu and Wratten, 2013).

The fourth and the current sustainability phase began in the 1980s and was the result of a growing concern for production efficiency, social responsibility, and the environment. A much broader base of stakeholders included corporations, nongovernment organizations (NGOs), retailers, and consumers as society began to view food and agricultural production in a more holistic manner. New intersector relationships developed and a realignment of organizational goals occurred. In the United States, the United States Department of Agriculture (USDA) refocused and renamed the Soil Conservation Service the Natural Resources Conservation Service (NRCS) to reflect this shift in focus from on-farm production and conservation needs to accounting for off-farm environmental impacts.

1.2   *A paradigm shift*

This seemingly simple shift in focus from on-farm needs to off-farm impacts continues to have profound implications for agriculture, governments, the environment, society, and the food system. It shifted agriculture conservation from largely the technical problem of solving erosion to an environmental problem with political, social, and economic dimensions. This shift in both scope and scale is comparable to going from "fixing the gully in the back 40," a relatively simple technical issue to solve, to "fixing the Gulf of Mexico hypoxic zone of the lower 48," a highly complex, socioeconomic, and ecological problem. The knowledge, skills, and abilities needed to address this new paradigm transcended the capacity of any single organization and sector, government or otherwise.

In addition to the vastly different scope and scale of the issue, new sustainability stakeholders emerged from corporate, NGOs, utilities, and financial sectors. The landscape is beginning to be viewed as *natural capital*, and recognized as the foundation on which our economies, societies, and prosperity are built (Rapacioli et al., 2014). These elements of nature, such as forests, rivers, land, minerals, and oceans, provide, in addition to the consumable goods, *ecosystem services*, such as carbon sequestration, water purification, biodiversity, and habitat and soil health. These services are now recognized as vital processes that are directly impacted by human activities.

In total, this seemingly simple shift in focus from on-farm needs to off-farm impacts exponentially increased the complexity of agricultural landscape sustainability by adding new values, stakeholders, organizations, and objectives. The problem was no longer a one-time technical issue to be addressed by conservation engineers, but instead became an ongoing socioeconomic problem within the context of local and global environmental issues. It became a *wicked* problem.

1.3   A wicked problem

Rittel and Webber (1973) introduced the term *wicked problems* to describe problems of an open societal system consisting of many variables and stakeholders. They are unlike ordinary or *tame* problems that can be solved with the traditional professions in science and engineering. Tame problems, such as landing a human on the moon, may be complicated, but are solvable and replicable. Wicked problems, such as raising a child, managing an economy, or addressing watershed concerns, are ongoing issues with a continuous one-shot opportunity to influence the outcomes. Rittel and Webber (1973) list 10 properties of wicked problems (Box 1.1) and compared them with ordinary or tame problems.

BOX 1.1   THE 10 PROPERTIES OF WICKED PROBLEMS

1. **There is no definitive formulation of a wicked problem.** It's not possible to write a well-defined statement of the problem, as can be done with an ordinary problem.
2. **Wicked problems have no stopping rule.** You can tell when you've reached a solution with an ordinary problem. With a wicked problem, the search for solutions never stops.
3. **Solutions to wicked problems are not true or false, but good or bad.** Ordinary problems have solutions that can be objectively evaluated as right or wrong. Choosing a solution to a wicked problem is largely a matter of judgment.

(Continued)

4. **There is no immediate and no ultimate test of a solution to a wicked problem.** It's possible to determine right away if a solution to an ordinary problem is working. But solutions to wicked problems generate unexpected consequences over time, making it difficult to measure their effectiveness.
5. **Every solution to a wicked problem is a "one-shot" operation; because there is no opportunity to learn by trial and error, every attempt counts significantly.** Solutions to ordinary problems can be easily tried and abandoned. With wicked problems, every implemented solution has consequences that cannot be undone.
6. **Wicked problems do not have an exhaustively describable set of potential solutions, nor is there a well-described set of permissible operations that may be incorporated into the plan.** Ordinary problems come with a limited set of potential solutions, by contrast.
7. **Every wicked problem is essentially unique.** An ordinary problem belongs to a class of similar problems that are all solved in the same way. A wicked problem is substantially without precedent; experience does not help you address it.
8. **Every wicked problem can be considered to be a symptom of another problem.** While an ordinary problem is self-contained, a wicked problem is entwined with other problems. However, those problems don't have one root cause.
9. **The existence of a discrepancy representing a wicked problem can be explained in numerous ways.** A wicked problem involves many stakeholders, who all will have different ideas about what the problem really is and what its causes are.
10. **The planner has no right to be wrong.** Problem solvers dealing with a wicked issue are held liable for the consequences of any actions they take, because those actions may have such a large impact and are hard to justify.

Source: Adapted from Rittel, H. W. J. and Webber, M. M., Dilemmas in a General Theory of Planning, Institute of Urban and Regional Development, University of California, Berkeley, CA, 1973.

Although Rittel and Webber focused primarily on problems related to social policy, their ideas have been applied to a variety of problems that have intersecting economic and ecological dimensions (Hart and Bell, 2013).

Most sustainability issues can be seen as *wicked problems* as they deal with social complexity having decentralized control and intelligence. There is a growing recognition that conventional approaches are inadequate to resolve wicked problems (Hart and Bell, 2013) and the dominant institutions addressing them require a change in their approach (Van Zeijl-Rozema et al., 2008).

Institutional scholars are seeing a need for a paradigm shift in how researchers approach complex societal problems. They have recognized scientific knowledge is necessary but not sufficient to resolve wicked problems (Hart et al., 2015). This paradigm shift toward alternative frameworks to address new societal challenges has helped shape the field of sustainability science (Kates et al., 2001; Hart and Bell, 2013). Sustainability science is a transdisciplinary option that leads to the question of how sustainability science can achieve positive impacts for wicked problems (Vermeulen and Campbell, 2011).

1.4   Sustainability science

Sustainability science has emerged over the last two decades as a vibrant field of research and innovation. Like "agricultural science," sustainability science is a field defined by the problems it addresses rather than by the disciplines it employs (Clark, 2007; Kates, 2011). The central purpose of sustainability science is not just to analyze problems but to contribute to solving them (Hart and Bell, 2013).

Sustainability science creates new knowledge, coproduced through close collaboration between scholars and practitioners (Clark and Dickson, 2003). This broader understanding is gained from observations, monitoring, and place-based studies. It is intended to aspire practitioners and scientists to move scientific knowledge into action (Kates, 2010).

Kates et al. (2001) proposes a set of core questions (Box 1.2) to guide different sectors to work in concert along sustainability trajectories.

Hart and Bell (2013) recognize sustainability science as science that (1) is problem-driven and focused on deriving and testing solutions based on scientific knowledge, (2) emphasizes the dynamic, coupled interactions between natural and human systems, and (3) stresses active and ongoing engagement with diverse stakeholders.

Sustainability science requires new approaches to make decisions under a wide range of uncertainties in natural and socioeconomic systems. A broader use of networks that alllows scientists and practitioners to engage new expertise and knowledge is critically needed (Kates, 2011). The Proceeding of the National Academy of Sciences (PNAS) recognized this need and created a "room of its own" for the maturing field of sustainability science. This new section in PNAS is intended to transcend the foundational disciplines of geography, ecology, economics, physics, and political science and focus instead on understanding the complex dynamics that arise from the interactions between human and environmental systems (Clark, 2007).

> ## BOX 1.2 CORE QUESTIONS OF SUSTAINABILITY SCIENCE
>
> 1. How can the dynamic interactions between nature and society—including lags and inertia—be better incorporated in emerging models and conceptualizations that integrate the earth system, human development, and sustainability?
> 2. How are long-term trends in environment and development, including consumption and population, reshaping nature–society interactions in ways relevant to sustainability?
> 3. What determines the vulnerability or resilience of the nature–society system in particular kinds of places and for particular types of ecosystems and human livelihoods?
> 4. Can scientifically meaningful "limits" or "boundaries" be defined that would provide effective warning of conditions beyond which the nature–society systems incur a significantly increased risk of serious degradation?
> 5. What systems of incentive structures—including markets, rules, norms, and scientific information—can most effectively improve social capacity to guide interactions between nature and society toward more sustainable trajectories?
> 6. How can today's operational systems for monitoring and reporting on environmental and social conditions be integrated or extended to provide more useful guidance for efforts to navigate a transition toward sustainability?
> 7. How can today's relatively independent activities of research planning, monitoring, assessment, and decision support be better integrated into systems for adaptive management and societal learning?

This field seeks to facilitate what the National Research Council has called a "transition toward sustainability," improving society's capacity to meet the needs of a much larger but stabilizing human population and sustain the life-support systems of the planet (Clark, 2007). To meet this sustainability challenge, an equal focus is placed on how social change shapes the environment and how environmental change shapes society (Clark and Dickson, 2003).

Stock and Burton (2011) state sustainability science is not a science by any usual definition. By studying sustainability through multi-, inter-, and transdisciplinary approaches, they noted that scientists are yet to develop a set of principles by which knowledge and sustainability can be systematically built. Rather it consists of the plethora of ideas, sometimes in conflict on how to achieve a sustainable future.

Sustainability science research is attempting to understand how to manage specific places where multiple efforts interact with multiple life-support systems to meet multiple human needs in highly complex and often unexpected ways (Clark, 2007). It is from this highly complex and integrated place-based scenario of agricultural landscape sustainability that wicked problems emerge.

1.5   The wicked features of agriculture sustainability

Rittel and Webber's (1973) 10 properties of wicked problems illustrate a dynamic interaction of multiple parties attempting to achieve multiple and often conflicting objectives on an ongoing basis. The challenges of making the ecological–economic connections, coordinating governance, aligning disparate stakeholders, and accounting for natural capital outputs are cornerstones of a wicked problem.

1.5.1   Ecological–economic disconnections

Agriculture's fourth phase is about reconnecting traditional economic values with that of natural capital and the ecosystem services it provides. Hart and Bell (2013) state this kind of research requires extensive interactions and collaboration between economists and ecologists. In theory, researchers from these two fields should have a lot in common given that the names of the two disciplines share a common root. *Oikos*, derived from the Greek word meaning *house,* is the root word for both "ecology" and "economy." *Ecology*, from Greek *oikos* and *logy* (the study of), is defined as the scientific study of the distribution and abundance of resources and the interaction between organisms and their natural world. *Economy*, from Greek oikos and *nomos*, is defined as the study of how humans use, supply, and distribute the resources that are available to them (Barnhart, 1988). Both systems allocate resources through a complex network of relationships.

Indeed, while it sometimes appears that economists and ecologists occupy entirely different houses that may or may not be located on the same planet, progress has been made in recent years as the two fields worked together on systems for modeling challenges and integration of research themes such as ecosystem services (Hart and Bell, 2013).

1.5.2   Conflicting governance styles

Governance is an awkward term that has different meanings for different individuals and organizations and is applied in various ways (Ho et al., 2014). In its broadest sense, it refers to the various ways through which organizations are coordinated and how "things get done" as it relates to decisions that define expectations, grant power, and verify performance (Meuleman, 2008).

Governance associated with agriculture landscape sustainability is not a straightforward concept and can be seen as a collection of rules, stakeholder involvement, and processes toward achieving many goals (Van Zeijl-Rozema et al., 2008). From an organizational approach, each governance style has its own logic as it relates to relationships, decision making, and compliance (Roe, 2013). Governance comes in several styles consisting of top-down hierarchies, profit-motive market styles, and trust-based networks. Mixing the three governance styles can lead to conflicts, competition, and unsatisfactory outcomes (Jessop, 2002). Meuleman (2013) states there are three problems with the application of these governance styles. First, they can undermine each other. Second, each of them has shortcomings. Third, each has attractiveness to certain personalities, leading stakeholders to perceive their governance style is superior.

Traditional *governing* by governments is undermined by socially complex, wicked problems of sustainability. Unlike typical top-down hierarchical structures, sustainability *governance* is seen as a shared responsibility of representatives from the state, the market, and civil society. Sustainability governance requires an acceptance that there are different perspectives on the concept of sustainability and awareness that multiple modes of governance may be able to steer the process of sustainability. Governance actors may provide leadership from top-down hierarchical styles or bottom-up network approaches (Van Zeijl-Rozema et al., 2008).

1.5.3   Disparate stakeholder values

Wicked problems are social problems created by different stakeholder perspectives and values. The definitions of *agriculture sustainability* are as diverse as the perspectives from individuals, organizations, sectors, and political affiliation. This creates a scenario where there is no agreement on what the problem is and no "right" solutions.

1.5.4   Variable natural capital

Perhaps the most tangible feature of this wicked sustainability issue is the natural capital of the agricultural landscape. Natural capital, like other economic capitals, is the basis for production. It is the soil, topography, plants, animals, and climate associated with the landscape. The less tangible features of natural capital include the bio-geo-chemical processes, functions, and interrelationships of the landscape that enable the agroecosystems to function and produce food and fiber commodities, and ecosystem services. How these components interact with each other and affect local, regional, and global ecosystems is very complex.

1.6   Putting sustainability science to practice

A challenge to putting sustainability science to practice is to determine
where to begin. Natural capital, disparate stakeholders, sustainability
governance, and the economical–ecological connections seem unwieldy
individually, let alone together. Rittel and Webber (1973) state there is
no formulation of a problem statement, no immediate test for a solution,
no signal when the problem is solved, and each attempt is a "one-shot"
proposal that may change the dynamics of the original wicked problem.
When addressing complex socioeconomic systems, Ostrom (2007) states
that there is no ideal entry point for carrying out rigorous, useful research
and the entry point depends on the question of major interest to the
researcher, user, or policy maker.

But one must begin. One strategy is based on Hart and Bell's (2013)
perspective that the heart of sustainability science is the interactions
between nature and society with stakeholder engagement as the corner-
stone. They view the practice of sustainability science to be a problem-
driven, solution-oriented science with the central purpose of not just to
analyze problems, but to contribute to solving them. They also stress the
importance of an economist's perspective. In short, we see a mutually beneficial research setting in which economists have "wicked good" training
to take on wicked problems and sustainability science has much to offer
to economists (Hart and Bell, 2013).

One potential focal point for stakeholder engagement, natural capital
valuation, sustainability governance, and making the ecological–economical
connection is through an *environmental market signal* (Gieseke, 2011).

1.6.1   An environmental market signal

An environmental market signal is a price or value signal for a particular environmental benefit. A traditional agricultural market signal would
be the price offered for a kilogram (or bushel) of grain. An environmental market signal is based on the price or value offered to produce an
environmental benefit. These environmental benefits are measured with
landscape indices representing the [ecosystem] services provided by the
landscape. A water quality index (WQI) represents the quality and/or
quantity of water that is shed by a field, farm, or other geographic area
with its numerical value-dependent landscape features and management
strategies.

One advantage of an index-based market signal is its capacity to communicate with many stakeholders on a variety of sustainability issues.
The efficiency of this process approaches economist Hayek's market *marvel* of price. Hayek called the free market system a marvel because just one
indicator, price, spontaneously carries so much information that it guides

buyers and sellers to make decisions. Like price that reflects thousands, even millions, of decisions made by people who do not know each other (Gwartney, 2010), a landscape index reflects various conditions and activities conducted in time and space that represent a level of sustainability. An index-based environmental market signal has the potential to carry the product or service's environmental information along with the price carrying the economic information.

1.6.2   *Spatially based trading platform*

Unlike the more tangible agriculture commodities such as grain and forages, the value of ecosystem services are often spatially and temporally based; that is, where and when they are produced may be as important as what was produced. While a bushel of wheat provides a consistent value no matter where it is, the value of a WQI is dependent on where the cleaner water is produced and when. A sustainability science approach would combine an environmental market signal with a geographically based platform that accounts for location and timing of the ecosystem service provided.

1.7   *Path to agricultural landscape sustainability*

Eleven case studies were analyzed for various strategies used to achieve their agricultural landscape sustainability objectives. Each was assessed for how agriculture sustainability was defined, accounted for, and valued and how they engaged stakeholders and organizations to produce the environmental benefits. The case studies, as an aggregate, revealed three primary principles of wicked problems and the corresponding components of a wicked solution. This dual finding is expected according to Rittel and Webber (1973), who state that to find the problem is the same thing as finding the solution, and that the problem cannot be defined until the solution has been found. They conclude that the process of formulating the problem and conceiving a solution is identical.

1.7.1   *Assessing wicked problems*

In Chapter 2, *An Enduring Wicked Problem*, three sources of the wicked agriculture sustainability problem are introduced as (1) the varied scope and scale of natural capital outputs, (2) the growing number of disparate agricultural stakeholders, and (3) conflicting governance styles of stakeholder organizations. Chapters 3 through 5 discuss the current status and provide examples for each problem source and how these three factors, individually and in concert, generate inefficiencies, confusion, and conflicts across organizations and sectors.

In brief, a wicked problem emerges as multiple stakeholders' demand and/or value different outcomes from the landscape. For example, grain production, water quality, and soil carbon sequestration may each be demanded from different entities at different scales. Demands for low-cost grain from a local processor, water quality at a national scale, and carbon sequestration at a global scale result in different market signals for different landscape outputs, with them often competing and conflicting. These outputs may be generated by a variety of simple and inexpensive activities or a series of very complex and expensive practices. The potential number of landscape patterns to produce the quantities and qualities of all these outputs in any given year is astronomical. Identifying the ideal landscape patterns that produce abundant food and cleaner water is overwhelming for any single entity (Ruhl et al., 2007).

1.7.2 Devising wicked solutions

In Chapter 6, *Devising a Wicked Solution*, three strategic actions to enable a wicked solution are introduced: (1) create a common landscape language, (2) align stakeholder sustainability activities, and (3) develop a shared governance platform.

The solution begins by organizing a landscape *language* by using vetted agroecological indices developed by government, land grant universities, and institutions. A significant cache of indices exist, but they must be scaled and harmonized to some degree to enable an alignment of stakeholder activities. This alignment is aided with the use of a *governance compass* to provide a new context for the interrelationships of potential stakeholders. The governance compass also describes stakeholders as *governance actors* representing one of four governance sectors: public policy maker, private policy maker, public practitioner, and private practitioner.

These two components (an index-based accounting system and the governance compass recognizing governance actors) enable a shared governance strategy, a concept based on the principles of creating partnerships, equity, accountability, and ownership at the point of services (Porter-O'Grady, 2001). It is at this point of service; the landscape, that activities are often initiated based on the presence or absence of a market signal. Chapters 6 through 9 discuss the accounting methods and governance styles of various case studies. Using these two traits (accounting method and governance style), a *governance footprint* was calculated for each of the 11 case studies. These case studies revealed organization governance styles trending away from top-down hierarchical governance toward more inclusive market and network governance. There was also an accompanied shift from practice-based accounting to outcome-based accounting. These more open governance and inclusive accounting systems enable stakeholders to value ecosystem services by using sustainability portfolios. These portfolios represent sustainability values beginning at the landscape and through corporate supply chains, government programs, and utilities.

1.7.3 Recognizing the glocal commons

In Section III, *Designing a Glocal [Business] Ecosystem*, the concept of glocalization is applied to the commons, governance, and the accounting of ecosystem services. *Glocalization* is a neologism meaning the combination of intense local and extensive global interaction (Wellman, 2002). *Glocalization* occurs when local actors have a more pronounced role in addressing global challenges. It denotes a merging of global opportunities and local interests to create a more socioeconomically balanced world (Roldan, 2011) and results in a combination of globalization and localization processes (Vries, 2010) and the *simultaneity* of globalizing and localizing processes (Blatter, 2007). This simultaneity of agriculture's impact results from it being a direct employer of 1.3 billion people with a footprint covering nearly 30% of the earth's land surface. The nature of agriculture, as an industry, affects local, regional, and global economies and ecosystems, simultaneously. The glocal phenomenon needs to be represented in sustainability governance, accounting systems, and commerce.

1.7.3.1 Global perspectives

At the global level, agriculture sustainability strategies and natural capital accounting are being sought by global financial, governmental, and corporate interests. A brief list of many similar efforts includes the following:

- The Natural Capital Declaration (NCD) is a global finance-led initiative to integrate natural capital considerations into financial products and services, and to work toward their inclusion in financial accounting, disclosure, and reporting. It is signed by the CEOs of 40 financial institutions, with supporting (nonfinancial) organizations (NCD, 2014).
- Wealth Accounting and the Valuation of Ecosystem Services (WAVES) is a World Bank–led global partnership that aims to promote sustainable development by ensuring that natural resources are mainstreamed in development planning and national economic accounts. This global partnership brings together a broad coalition of United Nations agencies, governments, international institutes, nongovernmental organizations, and academics to implement natural capital accounting where there are internationally agreed standards, and develop approaches for other ecosystem service accounts (Smith, 2014).

- The Natural Capital Coalition (NCC) is a global, multistakeholder open source platform for supporting the development of methods for natural and social capital valuation in business. The coalition brings together global stakeholders to study and standardize methods for natural capital accounting to enable its valuation and reporting in business (NCC, 2015).

1.7.3.2 Local and regional perspectives

At the local and regional levels, a listing of all efforts seeking a natural capital valuation strategy would number in the thousands. Several case studies discussed to various degrees include local, state, and regional efforts such as the following:

- Seattle Public Works and the Government Accounting Standards Board (GASB) on valuing the natural capital of a drinking water supply watershed
- The Sustainability Consortium led by corporate retailers to account for and improve the sustainability of supply chains
- A Minnesota state-level water quality certainty programs led by state and federal governments to account for agricultural practices to improve watersheds
- The Electric Power Research Institute leading a regional water quality trading program to account for conservation practices traded with waste water treatment facilities

These case studies represent efforts at the local, regional, and national levels and within the sectors of government, corporations, utilities, non-profit organizations, and agribusiness suppliers. In part, they were chosen to illustrate how disparate organizations and sectors have common sustainability objectives, but varied approaches to account for and value sustainability outcomes. In some instances, the same sustainability objective from the same parcel of land is sought, yet the processes are not integrated. This uncoordinated approach results in low and diffuse values for ecosystem services and creates high and multiple transaction costs. Within this context, it is not surprising that markets are unable to emerge and sustain. New values can be realized by reexamining and harnessing system complexity (Axelrod and Cohen, 1999).

1.8 The grand economic challenge

The demand for ecosystem services from farm and agriculture landscapes has grown dramatically—for more types of services and from multiple sectors. This increase is reflected in new government conservation programs, pollution regulations, sustainability standards in corporate supply chains, agriculture-related programs of nonprofit organizations, and the demand from consumers and the food industry for the eco-labeling of agricultural products (Sampson et al., 2013). The number and types of projects and efforts to account for and value ecosystem services are occurring at all scales.

The challenge is not just how to create demand for ecosystem services, but how to convert the existing demand from *absolute demand* to *effectual demand*. Adam Smith observed these differences in the relatively immature economic system of the eighteenth century. He recognized an *absolute demand* is the total demand for a commodity or product from all those who desire it whether or not they have the financial means to purchase it. An *effectual demand* is created when a sufficient price is applied toward a commodity to entice entrepreneurs to produce it (Smith, 1952). Today, these concepts apply to natural capital and ecosystem services. The absolute demand is steadily increasing, but an insufficient effectual demand exists due to the market development challenges of public goods.

The grand economic challenge is to capture and harness the glocal complexity of agriculture landscape sustainability. This includes a transaction process that supports sharing in the costs and values of agricultural landscape sustainability. It is viewing the transaction process from the lens of an ecological *supply web* rather than just the lens of an economic *supply chain*. It is about creating the capacity for stakeholders to interdependently value sustainability.

The Prince of Wales, speaking at The Prince's Accounting for Sustainability Forum at St. James's Palace in London, December 2013 (The Guardian, 2013) expressed these sentiments in this manner:

> In stark financial terms, all the evidence demonstrates a simple fact: we are failing to run the global bank that we call our planet in a competent manner. We no longer just take a dividend each year; instead, for some time, we have been digging deep into our capital reserves. And, after the near collapse of our entire financial system, we all know that such excessive risk-taking can cause immense havoc. The ultimate bank on which we all depend—the bank of natural capital—is in the red; the debt is getting ever bigger and that is reducing Nature's resilience and considerably impeding her ability to restock. It leaves us dangerously exposed.

While addressing natural capital accounting seemed impossible a decade ago, it is now becoming possible by a shift in social values, access to big data, an interconnected society, cryptocurrencies, and the emergence of multisided business platforms (MSPs). MSPs are unique business structures that are disrupting traditional business models in a big way. Goodwin (2015), senior vice president of strategy and innovation at Havas Media, stated

> Uber, the world's largest taxi company, owns no vehicles. Facebook, the world's most popular media owner, creates no content. Alibaba, the most valuable retailer, has no inventory. And Airbnb, the world's largest accommodation provider, owns no real estate. Something special is happening.

What appears to be happening is people are imagining new ways to solve problems with the use of big data and a networked society. The MSPs are capturing existing *unvalued* values of the economy. In the case of Uber, it is unused cars and with Airbnb it is unused living quarters. Even more interesting is Alibaba's creation of an e-commerce ecosystem. Alibaba is generating wealth, not by selling goods, but by creating the structure, processes, and functions of an e-commerce ecosystem and harnessing its complexity. Alibaba, in many respects, is enabling people and businesses to generate new values through the creation of *e-commerce ecosystem services.*

By venturing down these platform and ecosystem paths, these entities are providing a wicked solution to a wicked problem few knew existed. Few recognized the *problem* of unused cars and unused living quarters. Few recognized there were problems in commerce preventing millions of people from interacting and generating transactions. Perhaps, as Rittel and Webber (1973) predicted, they found the problem by finding the solution. As they provided new ways to engage people, it was recognized that the *old way* of doing things was the *wrong way* relative to value creation.

This book proposes the development of a multisided shared governance (MSSG) platform to support an eco-commerce ecosystem. As a geographical-based platform, it consists of delineating the landscape into assessable units with the capacity to calculate ecosystem services: the natural capital outputs and outcomes of the agricultural landscape. The MSSG platform enables sustainability stakeholders to interact at the interface of the two very complex systems of agriculture and the economy to account for and exchange values. This new eco-commerce system is possible by enabling engagement via shared governance; bringing partnerships, equity, accountability, and ownership at the point of service; at the landscape.

And ultimately, the success of the eco-commerce ecosystem is dependent on a design that recognizes that the economic system is a dependent subsystem of the earth's ecological system.

section one

An enduring wicked problem

chapter two

An enduring wicked problem

Finding the path toward agricultural landscape sustainability remains a significant challenge in the early twenty-first century. This challenge of reaping economic rewards while sustaining the landscape has long been with humankind. Made famous by Garrett Hardin's (1968) *Tragedy of the Commons* essay and noted nearly 2500 years earlier in the writings of Aristotle, society's answer to landscape sustainability remains unresolved (Waldron, 2004).

2.1 A wicked problem

Achieving landscape sustainability is a complex endeavor consisting of social, economic, and ecological components. The agricultural landscape has functional qualities at many scales and scopes. From the incalculable biodiversity interworking of *a teaspoon of soil* to the water purification processes within a great river basin, each is interrelated and undefinable. Within these contexts, there are many diverse stakeholders having common and conflicting objectives. Organizations, numbering in the thousands, are involved in all aspects of landscape use and sustainability. By many accounts, achieving agricultural landscape sustainability is a *wicked problem* (Balint, 2011; Bruggemann et al., 2012; Davies, 2015; Howes and Wyrwoll, 2012; Jones, 2011; Hearnshaw et al., 2011; Hunt and Thornsbury, 2014; Ostrom, 2007; Sørensen and Torfing, 2012; Scholz and Stiftel, 2005).

The term *wicked* in this context is used, not in the sense of evil, but rather as an issue highly resistant to resolution (Commonwealth of Australia, 2007). Since wicked problems are not always clear in their solution or in the steps to be taken, it is difficult to know whether the desired outcomes are being produced or even ultimately achieved. Wicked problems have become commonplace as these types of problems include nearly all social issues involving human health, education, transportation, and the environment. Wicked problems are a product of the increasing complexity and uncertainty of the physical world and social issues (Meuleman, 2013).

In contrast, *tame* problems are those that scientists and engineers usually work on (Rittel and Webber, 1973). For example, landing a human on the moon, sequencing DNA, and building a dam are tame problems. For each, the task is relatively clear; specific steps can be made and then

Values Knowledge	Consensus	Disagreement
Consensus	Technical	Political
Disagreement	Scientific	Wicked

Figure 2.1 The types and sources of problems are related to the consensus and disagreement on the knowledge and values associated with an issue. If there is consensus on both knowledge and values the problem to be resolved is technical in nature. Scientific and political problems are based on different combinations of consensus and disagreement on the values and knowledge of an issue. The source of a wicked problem is disagreement on both values and knowledge of an issue. (From Meuleman, L., *Transgovernance: Advancing Sustainability Governance,* Springer, Heidelberg, 2013.)

replicated to achieve the outcome again. Tame problems, although often complicated, are relatively straightforward as to whether or not the problems have been solved.

Generally, the source of wicked problems is rooted in the lack of consensus as it relates to the values and knowledge of an issue (Meuleman, 2013). Figure 2.1 describes the type of issue to be addressed when there are various combinations of consensus and disagreement on values and/or knowledge. When consensus occurs on values and knowledge, the issues to be resolved are technical. When consensus occurs on values, but not on knowledge, the issues to be resolved are scientific. When consensus occurs on knowledge, but not on values, the issue to be resolved is political. All of these are tame by nature. When there is disagreement on values and knowledge, this issue is wicked.

Wicked problems arise from the interaction of multiple stakeholders with incomplete knowledge and conflicting preferences for the issue at hand. Wicked problem complexity is twofold: complexity of the issue or system and the differing preferences of stakeholders. This twofold complexity renders wicked problems irreducible (Hearnshaw et al., 2011). Socioecological systems such as agricultural landscape sustainability are wicked issues.

2.1.1   Sources of wicked (landscape sustainability) problems

The source of wicked problems cannot be found by traditional problem-solving tactics (Rittel and Webber, 1973). Pure study amounts to procrastination, because little can be learned about a wicked problem by objective data gathering and analysis. Wicked problems demand an opportunity-driven approach: they require making decisions, doing experiments, launching pilot programs, and testing prototypes (Christensen, 2009).

To identify specific problems embedded in socioecological systems, Ostrom (2007) created a diagnostic framework consisting of the system attributes. By focusing on the attributes of (1) the resource system, (2) the resource outputs, (3) the users of the system, and (4) the governance of the system, it gave insights into how the attributes may affect or be affected by socioeconomic, political, and ecological settings in which they are embedded. For example, the issues related to the sustainability of a pasture (resource system) could be diagnosed by measuring the quantity of fodder it produced (resource output) and determining the number and extent of cattle grazing (users) based on the formal and informal rules (governance) of use. Within this framework, Ostrom was able to postulate how changes in one attribute may affect the socioecological system overall.

Jones (2011), a colleague of Elinor Ostrom and researcher of complex social problems, identified three sources of wicked, complex problems. First, the capacities to tackle complex problems are often distributed across a range of stakeholders. Second, complex problems are difficult to predict. And third, complex problems involve conflicting goals. The combination of these sources causes wicked problems to manifest themselves in different ways at different levels. Decision makers at one level perceive solutions differently than those at other levels. The result may be many divergent but equally plausible interpretations of the issue.

Three sources of the wicked problem associated with agriculture landscape sustainability were identified by analyzing 11 case studies. The sources originated within the complexity of the physical environment and social institutions. The following three sources were identified:

1. The varied scope and scale of natural capital outputs and outcomes
2. A growing number of disparate stakeholder values and accounting systems
3. Conflicting governance styles of stakeholder organizations

Individually, each of these problems present challenges to accounting for and valuing the benefits derived from agriculture landscapes. When combined, the complexity of the problem significantly increases. The varied landscape outputs consist of crops, livestock, timber, water, carbon, habitat, and many other benefits. These are produced daily, seasonally, yearly, and decadal, and are sought by numerous stakeholders with different values and intentions. These stakeholders represent hundreds of organizations with differing and often conflicting governance styles. The interrelationship of these three principle problems compounds the complexity of sustaining agricultural landscapes and creates an enduring wicked problem.

2.1.1.1 Varied scope and scale of natural capital outputs and outcomes

Natural capital is one of five generally accepted economic capitals identified along with financial, physical, human, and social. Natural capital, like other economic capitals, refers to "factors of production" that are used to create goods and services, but are not themselves directly consumed. Natural capital is the stock of natural ecosystems that yields a flow of valuable ecosystem goods and services. Ecosystem goods are defined as the tangible consumable goods such as grains, fiber, and water that are produced from the landscape. Ecosystem services are the less tangible processes of the landscape that purify water, pollinate crops, and support wildlife (Daily, 1997).

The scope and scale of agriculture landscape outputs vary greatly. The output can be a specific good, such as a crop at the field scale or water at watershed scale. It may be a specific service, such as pollination, that supports a regional ecosystem or a service, such as carbon sequestration, that affects the atmosphere at the global scale. Despite the great variations in scope and scale, these natural capital components are related ecologically and often related economically. Because of this interdependency, the need to measure, account for, and incorporate these varied values into the economic system is becoming a necessity.

The first-order complexity of the physical nature of agriculture sustainability is compounded by various sectors of agriculture, forestry, fisheries, industries, corporations, governments, NGOs, finance, consumers, and citizens, with each having unique perspectives and demands from the landscape.

2.1.1.2 Growing number of disparate stakeholder values

Prior to the 1970s, agriculture stakeholders in the United States represented local, state, and federal agencies, land grant universities, land owners, farmers, agribusiness retailers, and food processors. Since the 1970s, a growing awareness in the environment and the value of natural capital has greatly expanded the number of stakeholders interested in the agriculture landscape. Stakeholders now include groups associated with bio-energy, environmental advocacy, animal welfare, consumers, corporations, trade groups, private research, and food retail (Sampson et al. 2013).

In addition to stakeholder groups, individual citizens and consumers are demanding more transparency from food processors as it relates to food ingredients and the methods used to grow, process, and transport foods. Corporations, in turn, are applying procurement criteria to suppliers and growers to assess the sustainability of products (The Sustainability Consortium, 2015).

This second-order level of complexity is further compounded by the stakeholders' unique organizational *governance style*. Governance styles can be defined as the processes of decision making and implementation, including the manner in which the organizations relate to each other (Kersbergen and Waarden, 2004).

2.1.1.3 Conflicting governance styles

Governance is an awkward concept as it has different meanings for different individuals and organizations and is applied in various ways (Ho et al., 2014). Governance is a broader term than government, as it is a fundamental component of political, business, social, and government organizations. In its broadest sense, it refers to the various ways through which society is coordinated. Therefore, it is possible to have governance without government or government can be seen as just one of the institutions involved in governance (Rhodes, 1996).

Governance relates to decisions that define expectations, grant power, and verify performance. Governance is also the process of how organizations acquire knowledge and apply strategies (Nickerson and Zenger, 2004). Simply put, governance is how an organization "gets things done" (Meuleman, 2008).

Organizations rely on three governance styles (Thompson, 1991):

1. Hierarchy: a conformist top-down framework with emphasis placed on seniority. Militaries, schools, governments, and churches are often associated with a hierarchy model.
2. Market: an incentive-based top-down framework with emphasis placed on innovation and profit incentives. Corporations are often associated with a market model.
3. Network: a nonhierarchical framework with emphasis placed on purpose and trust. NGOs are often associated with a network model.

Figure 2.2 illustrates each of these governance styles in a schematic form. Hierarchy governance promotes conformity and subordinate accountability. Market governance relies on top-down control, but promotes innovation and profiteering over conformity. Network governance is formed around multiple relationships in both horizontal and vertical manners.

Conflicts arise from the sets of values each of these governance styles contain. Hierarchy governance values are based on the expectation that there should be a *subordinate* to the hierarchy and it relies on regulations and control instruments to meet goals. Market-based governance values a *customer* perspective and relies on competition and innovation to

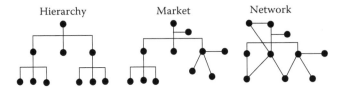

Figure 2.2 Organization governance is based on one of three typical governance styles. Hierarchy is top-down governance based on conformance, market is top-down governance based on incentives, and network governance is based on interconnected relationships.

achieve results. Network-based governance seeks *partners and cocreators* and relies on trust to achieve outcomes (Meuleman, 2008). Mixing these varied values and expectations, even when addressing common objectives, generates conflicts among stakeholders. These three orders of complexity based on natural capital, the varied social and economic values of stakeholders, and their organizational structures combine to create a complex agriculture landscape system.

2.2   The complex agriculture landscape system

The agriculture system of today emerged as complex systems do, from the interactions of all the system components and stakeholders (Mitchell and Newman, 2002). As a complex system, it is an ensemble of many elements which are interacting in a disordered way, resulting in robust organization and system memory (Ladyman and Lambert, 2012). Like most all complex systems, it is a composition of simple, complicated, and complex subsystems.

2.2.1   The simple and complicated

Simple and complicated systems are manageable with central control and their outcomes can be predetermined. For example, a simple task such as turning a key to start a tractor is a simple system task. Using the tractor to plant a crop or construct a waterway in a field is a complicated system task. It is a complicated process based on a variety of scientific knowledge sets.

In a simple system, a process can be followed and repeated with relatively little expertise and be expected to produce uniform results. In complicated systems, a higher order of expertise is often required and a variety of disciplines or expertise may need to be drawn upon in order to produce a successful result. Once that result is achieved, it is—in most cases— replicable (Glouberman and Zimmerman, 2002). With time, experience, and repetition, planting a crop or constructing a waterway and knowing when the task is successfully completed is knowable. Addressing a simple and complicated system is considered a tame problem.

2.2.2   The complex

Complex systems differ from simple and complicated systems. The distinction between complicated and complex systems is important in evaluating problems as complicated and complex systems require different methods of analysis. Good evaluation of a complicated system involves repetition, replication, predictability, and infinite detail. Good evaluation of a complex system involves pattern description, contextualization, and dynamic evolution (Williams and Van 't Hof, 2014).

In a complex system, outcomes are an emergent quality of the system which can be influenced, but not controlled, by stakeholders (Ng and Andreu, 2012). Many unknowns remain and the systems are in a state of constant flux and unpredictability. There are no right answers, only emergent behaviors (Snyder, 2013).

In complex systems, the individual elements are influenced directly by the behavior of the system as a whole, and at the same time their interactions lead to the emergent behavior at the aggregate level of the system. The *common sense* connection between the size of an event and its consequences no longer holds. Small changes have the capacity to trigger large-scale events (Farmer et al., 2012).

A complex system can only be understood as a whole; it is different from the sum of its parts and often involves a socioeconomic aspect. The system itself emerges from the interaction of the parts making it dynamic, probabilistic, and unpredictable (Williams and Van 't Hof, 2014; Farmer et al., 2012). To better understand complex, wicked problems and their components, Williams and Van 't Hof (2014) suggest drawing a rich picture (see Box 2.1) to define the complex system as it *is*.

BOX 2.1

Rich pictures are a part of soft system methodology, a systemic approach developed by Peter Checkland for tackling real-world problematic situations, the kind of messy problem situations that lack a formal problem definition.

The rich pictures provide a mechanism for learning about complex or ill-defined problems by drawing detailed ("rich") representations of them. Typically, rich pictures consist of symbols, sketches, or "doodles" and can contain as much (pictorial) information as is deemed necessary (Avison et al., 1992).

2.2.3   The system as it is

Rittel and Webber (1973) describe wicked problems as discrepancies between the state of affairs as it "is" and the state as it "ought to be." To begin a problem-solving effort one must create a representation of the problem to make the solution transparent (Simon, 1996) or help illuminate important dimensions of a problem (Baldwin and Woodard, 2009). In this case, the primary dimensions include the natural capital, stakeholders, and organizational governance.

Drawing a rich picture is one method to identify components, stakeholders, and boundaries of the [agricultural sustainability] system as it is (Simon and Bell, 2008; Williams and Van 't Hof, 2014). Since no endeavor is boundless, system boundaries must be consciously chosen to determine which components are included within the system. This process deems what is relevant, who is important, what processes are accounted

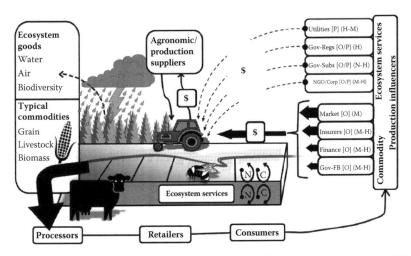

Figure 2.3 A *rich picture* is used to depict the boundaries and components of the agriculture landscape production system as it currently exists. The landscape and its features are the natural capital that produces ecosystem goods and ecosystem services. The typical food and fiber commodities are sold to processors and on to consumers. The activities of land managers directly influence the production of commodities and ecosystem goods and services. Indirect influence on commodity production is applied from government, financial institutions, insurers, and the market to produce commodities in a fairly organized and strong manner. Commodity production influencers are depicted as using outcome-based accounting and market-hierarchy governance styles. Indirect influence on ecosystem service production is applied from government, NGOs, corporations, and utilities in a less organized and relatively weaker manner. Ecosystem service production influencers generally use a mix of outcome and practice-based accounting systems and a mix of governance styles.

for, and what values are expressed (Williams and Van 't Hof, 2014). The system boundaries shown in Figure 2.3 are defined by the natural capital components, the stakeholders directly and indirectly influencing the land management decisions, and the governance styles adopted by the organizations.

The system as is contains the natural capital that produces ecosystem goods and ecosystem services (see Box 2.2). The system supply chain is represented by processors, retailers, and consumers that influence numerous sectors and organizations described here as commodity and ecosystem service production influencers. The commodity production influencers rely on outcome-based [O] accounting systems (direct measurement or indices) and market and hierarchy governance styles. The relatively uniform structure generates a concerted approach to valuing commodity production. The ecosystem service production influencers consist of government regulators and subsidy programs, utilities, NGOs, and corporations. They rely on outcome and practice-based [O/P] accounting systems and a mix of governance styles. This relatively nonuniformed structure results in low and diffuse values for ecosystem services. The production group consists of the farmers and businesses that directly support the production of both commodities and ecosystem services depending on the market and policy signals they respond to.

BOX 2.2   ECOSYSTEM GOODS AND ECOSYSTEM SERVICES

The definitions of ecosystem goods and ecosystem services vary widely and are discussed in more detail in Chapter 3. In this text, the definition of ecosystem goods and services are aligned with the traditional economic description where a service is an *intangible process* that cannot be weighed or measured, and a good is a *tangible output* of a process that has physical dimensions.

Ecosystem services are the less tangible functions and processes of natural capital that consisted of pollination services, nutrient and carbon cycling, and water purification. Ecosystem services are seldom included within economic transactions. "Ecosystem goods" is a broad term representing goods often not traded such as water, air, and biodiversity as well as typical commodities that enter the food supply chain.

With this definition, the water flowing from the landscape is an ecosystem good, whereas the landscape's process of cleansing the water is an ecosystem service. Likewise a bushel of grain is a good and the nutrient cycling of the soil and crop is an ecosystem service.

2.2.3.1   Natural capital

The natural capital of the agriculture landscape consists of soils, topography, biology, climate, and the overall functions of the agriculture ecosystem. The natural capital outputs (discussed in more detail in Chapter 3) are categorized as ecosystem goods and ecosystem services (Box 2.2). The rich picture depicts two types of ecosystem goods: (1) traditional commodities (grain, livestock, and biomass) that are processed into foods and products for consumers and (2) consumable, but generally noncommoditized goods such as water, air, and biodiversity that are vitally important to the economy, ecology, and human welfare. The flow of traditional commodities is depicted by the large downward facing arrow in the lower left portion and the flow of the consumable, but generally noncommoditized goods are depicted by the dashed arrow in the upper left portion.

The other category of natural capital outputs include ecosystem services, the processes and functions of the agriculture landscape that are rarely included within economic transactions. These include nutrient cycling, carbon sequestration, pollination, erosion control, water purification, and other bio-geo-chemical functions of a landscape. They are depicted as diffuse and distributed outcomes from the landscape as a whole.

The production of ecosystem goods and services is directly dependent on the management activities of the land managers whose decisions are influenced by numerous system stakeholders.

2.2.3.2   System stakeholders

System stakeholders consist of producers, processors and consumers, and corporate, government, NGO, and utility entities each with diverse sets of values and interest in agriculture landscape management.

Generally speaking, in the system as it *is*, there are well-established institutions, cultural norms, and values that support the production of traditional commodities. This group is represented in the lower right-hand portion of the figure and is referred to as the commodity production influencers. There are also growing, but far less established and organized group of institutions associated with supporting the production of ecosystem services. This group is represented in the upper right-hand portion of the figure and is referred to as the ecosystem service production influencers.

2.2.3.3   Production group

The production group is the land practitioners such as farmers and businesses directly involved in the production of ecosystem goods and services. Those in the production group are often described as *price-takers* and *policy-takers*, meaning they have little or no control on setting commodity prices and must work within the framework of government policies and the emerging corporate policies regarding sustainability practices. The producers' decisions are influenced by market price signals, risk factors and risk mitigation options, and government and corporate policies.

Ultimately, it is the actions of land managers, within the context of weather and environmental forces, which result in the production of various quantities and qualities of ecosystem goods and ecosystem services. Figure 2.3 also portrays the relative influence the production group receives from the production influencers. The values from the ecosystem service influencers are diffuse and low, and the values from the commodity influencers are combined and greater.

2.2.3.3.1 Production influencers Commodity and ecosystem service production influencers exert influence directly toward the production group. In reality, these two groups are not delineated as distinctly as illustrated, as their influence comes from a variety of sources with, at times, the intent to achieve both objectives. The processor–retailer–consumer grouping has a long-standing interest in abundant production and low food cost. Due to the relatively recent sustainability movement, many of the entities within this group now have sustainability interests as well.

Food processors provide a close link to commodity production by purchasing grains and livestock and generating current and future price levels or market signals. Consumers are adopting new values associated with food, causing retailers to seek ways to provide low-cost foods while considering their sustainability characteristics. This influence is passed on to commodity production influencers and ecosystem service production influencers.

2.2.3.3.2 Commodity production influencers The grain commodity market, crop insurers, production financers, and government subsidies provide a unified and additive approach in supporting the production of traditional commodities. Each sector in the production influencers group has a vested interest in generating maximum production of grain commodities. This common objective, over time, has generated complementary policies and strategies to support the production group in producing traditional commodities. This established and unified approach is illustrated by heavy and combined arrows leading to the production group.

Prior to the sustainability movement, the food processors, retailers, and consumers had little interest other than food price and so the commodity production influencers, for all intents and purposes, were the only influence the production group received.

2.2.3.3.2.1 Ecosystem service production influencers The ecosystem service production influencers, unlike the commodity production influencers, are rather diffuse in their influence as represented by dashed, unconnected lines. Of the four sectors represented, the Gov-Subsidy sector has applied influence the longest. The United States Department of

Agriculture (USDA) conservation incentive programs began in the 1930s with the creation of the Soil Conservation Service and the local conservation districts. In the 1970s, the Clean Water Act was passed and was followed by the sustainability movement of the 1980s.

This influence has continued to grow with support from NGO efforts to influence policy and social norms, and from corporate sustainability supply chain initiatives to source sustainably grown commodities. Generally, these stakeholders support land management strategies that generate cleaner water, improve soil health, sequester carbon, and support wildlife and other nonmarket goods and services related to the environment.

Today, the system as it *is* has a far greater number of sustainability stakeholders influencing the agriculture production system than decades ago, but the effects of the influence are mixed. This is due, in part, to the varied scope and scale of natural capital outputs and the challenges it causes for policy makers and market developers. These produce wicked challenges that overwhelm the efforts of the ecosystem service production influencers.

2.2.3.3.2.2 Seemingly infinite complexity Natural systems, such as agroecological systems, are not just a compilation of complicated subsystems, but are highly complex systems when viewed in terms of their number of constituents, the dynamic structure of their interconnections, the number of possible interactions, and the consequences of outside influences (Fisher, 2006). The intertwining of social, economic, and ecological aspects of agricultural landscape sustainability creates a scenario with seemingly infinite complexity.

To put this level of complexity into context, Ruhl et al. (2007) apply mathematics of permutations. To sustain an agricultural landscape, it must be managed in a way that produces abundant foods while maintaining the productivity of the natural capital. Ruhl et al. (2007) state that the number of potential land use patterns to accomplish this is related to the number of fields within a geographical area raised to the number of land-use options. For example, if a region contained 1000 fields and each producer has the option to plant five types of crops, the potential number of land-use patterns relative to crop production would be 1000 to the power of 5 or 10^{15}. Within this large number of possibilities, one landscape pattern emerges each year as farmers attempt to achieve their production goals within the context of market forces and government policies. While crop production does fluctuate each year, the market signals and government policies of the system as it *is* has been generally successful at creating a landscape pattern out of the 10^{15} possibilities to produce sufficient crops.

To put this scenario in the context of agricultural landscape sustainability, one could add several conservation practices and activities to the five crop options to calculate the number of potential landscape patterns within the context of a sustainable landscape. Incorporating five additional land management options into Ruhl's equation results in 10^{30} land-use pattern (1000 to the power of 10) possibilities. Obviously, many potential patterns would suffice for producing a sustainable agriculture landscape, but a process still needs to be in place to *choose* one. Within this socioeconomic context, it is apparent how the traditional top-down approach can be costly and be overwhelmed in the quest to determine a sufficient landscape pattern for these 1000 fields.

2.2.3.3.2.3 Undefinable property rights Unlike traditional commodity crops, ecosystem services do not have defined property rights. Without recognized property rights and access to markets, [natural capital] assets become "dead capital" unable to generate returns over and above that associated with their direct use. This ownership issue arises as ecosystem services have the characteristics of being defined as a *public good* rather than a *private good*. Public goods lack a clear definition and allocation of rights (Landell-Mills and Porras, 2002). Public goods are distinguished by their nonexcludability and nonrivalry status.

Nonexcludability means that consumers cannot be prevented from enjoying the good or service in question, even if they do not pay for the privilege. For instance, it is difficult, if not impossible, to exclude downstream communities from benefiting from improved water quality associated with sustainable practice upstream.

Where goods are nonrival, the consumption of a good or service by one individual does not reduce the amount available to others. In this situation, there is no competition in consumption since, theoretically, an infinite number of consumers can use the given quantity supplied. An example of a nonrival ecosystem service is carbon sequestration. Once carbon is sequestered, the global community may benefit from this in terms of a reduced threat of global warming.

In the system as it *is*, many ecosystem goods and services remain undefined with regard to property rights. And in one way or another, all environmental and natural resource problems associated with overexploitation or underprovision of public goods arise from incompletely defined and enforced property rights (Libecap, 2009).

2.2.3.3.2.4 Undermine market formation Where nonexcludability and nonrivalry exist, they undermine the formation of markets since beneficiaries of the good or services have no incentive to pay suppliers. As long as an individual cannot be excluded from using a good, they have little reason to pay for access. Similarly, where goods are nonrival, consumers know that where someone else pays, they will benefit. In both cases, beneficiaries plan to *free-ride* based on others' payments (Kim et al., 1980). And, where everyone adopts free-riding strategies, willingness to pay for public goods will be zero and will undermine markets.

2.2.3.3.2.5 High transaction costs   Trading in any market will not occur if the transaction costs exceed the benefits of a potential trade. Transaction costs include not only components of paperwork, verification, auditing, commissions, and other activities surrounding the sale of an item, but also the costs associated with the creation and operation of markets. Costs of market creation include defining property rights, setting up exchange systems, educating market participants, establishing monitoring and enforcement mechanisms, and building confidence in the system. Market operation includes costs of information gathering, negotiation, contract formulation, monitoring, and enforcement.

Ecosystem service credit-trading programs are characterized by higher transaction costs because trading partners may be widely distributed and there may be uncertainty in market components (Abdalla, 2008). The system as it *is* carries a high level of market uncertainty in measurement, valuation, verification, and defining property rights.

2.2.3.4   *Organizational governance styles*

The third wicked aspect of the system as it *is* involves the governance styles of the stakeholder organizations. Governance provides the framework for how an organization functions, gathers information, and makes decisions. In Figure 2.3, governance styles of production influencers are shown as well as how they account for the natural capital outputs and outcomes of interest.

The letters in the set of parentheses describe the primary governance style of the organizations. "H" is for hierarchy, "M" for market, and "N" for network governance styles. Since few organizations have strictly one type of governance style, the identifiers chosen are those that best represent the predominant governance style of the organization. Market and hierarchy governance are the predominant styles used by the commodity production influencers and a mix of governance styles is used by the ecosystem service production influencers.

Governance styles do influence which accounting strategies are adopted, as it relates to how information is gathered. That correlation is not made here, but accounting strategies are noted. The accounting strategy of the organizations are described in the first set of brackets ([ ]) and it includes an "O" for outcome-based, "P" for practice-based, or "O/P" to denote the use of the combination of the two accounting methods. For example, outcome-based accounting may be based on a direct measurement such as kg/hectare or a model calculation such as a water quality index that uses soil type, land slope, vegetation, climate, and other environmental and management factors to calculate the production of a particular quality and/or quality of water or other environmental benefit. A practice-based accounting system would identify particular practices, such as if a grass waterway is installed and functioning, or if a particular tillage practice is used.

Practice-based accounting is applied where measurements are difficult to obtain or due to traditional preference. Of course, the use of practice-based accounting, such as when a field was cultivated, would not be considered a viable option account for the production of grain where quantity and quality is easily measured and tested.

The commodity production influencers have adopted a relatively uniform outcome-based system using the common governance styles of market and hierarchy. The ecosystem service production influencers have adopted different accounting strategies and a mix of governance styles. This mix of accounting and governance strategies of the ecosystem service production influencers and the commodity production influencers is a fundamental aspect of the wicked problem.

The results are low and diffuse values, and high and multiple transaction costs for ecosystem services. This issue is compounded if one considers the ecosystem service production influencers must compete with the commodity production influencers, who over time have created multiple value streams for commodities and reduced the transaction costs associated with their markets.

2.3   Transdisciplinary challenge

The state of the system as it *is* consists of many organizations, sectors, and disciplines operating from isolated, silo-like perspectives. These traditional disciplinary approaches have had a considerable and overwhelmingly positive impact on the world and have provided scientists with methodological approaches and technologies (Brown, 2010). In addition, they provide shared concepts and language and accreditation to practitioners within their fields. Yet, it is increasingly recognized that, to address many of our current problems, this traditional approach is not sufficient (Stock and Burton, 2011).

The complexity of today's problems increasingly demand that scientists [and policy makers] move beyond the confines of their own discipline [and sectors], especially for problems concerning sustainability as natural, social, and human issues are tightly connected (Tappeiner et al., 2007). Initiating these changes and achieving sustainability objectives inherently requires transdisciplinary attempts (Stock and Burton, 2011).

Brown (2010) stated any resolution to socially embedded wicked problems calls for changes in that society in the form of new governance, and new methods of research and decision making based on transdisciplinary research. This does not entail asking researchers to reject the powerful tools and strategies developed by individual disciplines that led to disease reduction, an increase in world food production, and putting a man on the moon. It does ask for open transdisciplinary modes of inquiry that meet the needs of the individual, the community, the specialist traditions, and influential organizations. It requires an evolution in disciplinary thinking and imagination (Brown, 2010).

2.3.1 Disciplinary evolution

During the four centuries of the industrial revolution, analytic modes of scientific inquiry were developed to address technical problems and their social consequences. These became the academic disciplines that came to dominate educational institutions and research traditions (Brown, 2010). Hunt and Thornsbury (2014) describe this evolution from disciplinary to multidisciplinary, to interdisciplinary, and to transdisciplinary research. Figure 2.4 illustrates the evolutionary stages of disciplinary research. Choi and Pak (2006) define these disciplinary approaches with the common words of additive (multidisciplinary), interactive (interdisciplinary), and holistic (transdisciplinary).

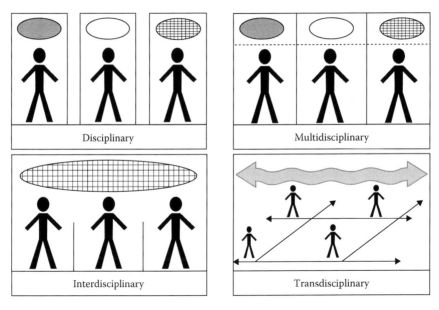

Figure 2.4 The four stages represent the evolution in disciplinary research. Disciplines acting singularly in their research create knowledge, theory, language, and boundaries for a particular area of study. As issues become more complex, multiple disciplines exchange knowledge with each other but do not integrate. Interdisciplinary research crosses disciplinary boundaries and results in the development of integrated knowledge and theory. Transdisciplinary research crosses disciplinary and academic boundaries with common goal setting. It is the most recent and challenging research method as it consists of academic and nonacademic participants addressing ongoing, real-world issues. This open-ended scenario is represented by an open-ended platform depicted as potentially unlimited and ongoing. (Adapted from Hunt, F. and Thornsbury, S., *Open J. Soc. Sci.*, 02(04), 340–351, 2014.)

A disciplinary approach involves goal-setting within one discipline with no cooperation or knowledge sharing. Its purpose is to develop new disciplinary knowledge and theory (Hunt and Thornsbury, 2014). A multidisciplinary approach is the most basic level of involvement. It refers to different disciplines working on a problem in parallel without challenging their disciplinary boundaries. It is taken to be a combination of specializations for a particular purpose (Brown, 2010). Figure 2.4 shows the first phase of disciplinary interaction. Goal-setting occurs under a common theme with loose cooperation to exchange knowledge.

An interdisciplinary approach brings about reciprocal interaction between disciplines and blurs the disciplinary boundaries in order to generate new common methodologies, perspectives, and knowledge (Choi and Pak, 2006). It is the common ground between two specializations that may develop into a discipline of its own, such as the convergence of biology and chemistry (Brown, 2010). Interdisciplinary is an integration of knowledge and theory (Hunt and Thornsbury, 2014).

A transdisciplinary approach involves scientists from different disciplines as well as nonscientists and other stakeholders and, through role release and expansion, transcends the disciplinary boundaries to look at the dynamics of whole systems in a holistic way (Choi and Pak, 2006). It crosses not only disciplinary, scientific, and academic boundaries, but integrates nonacademic participants to develop integrated knowledge among society (Hunt and Thornsbury, 2014). In this sense, transdisciplinary efforts are the highest form of integrated project, involving not only multiple disciplines, but also multiple *nonacademic* participants (e.g., land managers, user groups, the general public) in a manner that combines an interdisciplinary strategy with participatory approaches (Stock and Burton, 2011). Figure 2.4 illustrates a transdisciplinary approach as a broad platform with an ongoing solution resolution process that is open to all disciplines, social sectors, and entrepreneurs. This dynamic and open process is applicable to resolving the ongoing and open-ended wicked problems of agriculture landscape sustainability.

2.3.2 Application challenges

While transdisciplinary efforts are probably the most desirable, they are also the most difficult to obtain. Some researchers are skeptical about whether it can be achieved at all, causing Stock and Burton (2011) to refer to it as the holy grail of research, where achieving transdisciplinary landscape research is an exception and where even interdisciplinary solutions are seldom reached.

The review of transdisciplinary studies by Brandt et al. (2012) supported these findings and in particular the challenge of engaging nonacademic practitioners. In the peer-review published case studies, only 9 of the 236 studies (3.8%) involved practitioners in the entire transdisciplinary process from problem definition to implementation. Scholars involved in transdisciplinary projects must be willing to transcend and integrate disciplinary paradigms with the goal of finding a unity of knowledge beyond individual fields of study. As an evolving field of research with varied definitions and strategies, several challenges to applying transdisciplinary projects remain (Lang et al., 2012).

To overcome these challenges, Brandt et al. (2012) identified five key needs when undertaking transdisciplinary approaches to sustainability sciences:

1. *Establish coherent framing.* A lack of a shared framing for a particular issue results in scientists and practitioners addressing the issue with differing strategies and not producing robust knowledge to solve sustainability problems.
2. *Integration of methods.* Integrating different disciplinary methods and developing research methods that enable learning processes is crucial for an effective science–society interface.
3. *Research process and knowledge production.* Sustainability science research processes must cater to all participants to gather the knowledge related to the process phases of problem definition, analysis, and generation and application of solutions to real-world problems.
4. *Practitioners' engagement.* The link between practitioners and scientists is a crucial element for information exchange, collaboration, and practitioner empowerment to make decisions.
5. *Generating impact.* The challenge to intensively engage with practitioners at the scale needed to produce change tends to constrain the focus of transdisciplinary research to local or regional scales.

2.3.3   A wicked resilient problem

The long enduring system as it *is* is resilient while harboring three wicked principle problems of landscape sustainability:

1. Outputs and outcomes of the system vary in scope and scale.
2. Stakeholders have disparate values of and accounting processes for the outputs and outcomes of the system.
3. Stakeholder organizations within the system have inherently conflicting governance styles.

Each of the generic principles of themselves produces challenges to resolve any problem and together they create a wicked socioecological issue. To resolve this, one must understand the fundamental principles of these wicked problems to enable a transdisciplinary approach that transcends each of the principle causes.

chapter three

Natural capital outputs and outcomes

The wicked challenges associated with measuring and accounting for natural capital outputs and outcomes is, in part, founded in the historic and limited acceptance of natural capital as an economic capital (Bennun et al., 2014). Harrison (1989) states this basic difference is that environmentalists regard natural resources as assets analogous to man-made capital, whereas they are treated as free gifts of nature in the national economic accounts. Boyd and Banzhaf (2007) add that the lack of a unified definition for ecosystem services is a significant reason for their exclusion. This has resulted in multiple definitions from the scientific community on what an ecosystem service *is*, which undoubtedly leads to confusion in determining how to account for and incorporate their value into the economic system.

3.1   A natural economic capital

Wealth is created through factors of production, or so-called capitals that are used to create goods or services. The list can vary, but the following represent five major economical capitals:

- Financial capital: money used by entrepreneurs and businesses to make products or provide services
- Infrastructural or physical capital: factories and other man-made enterprises used as a means of production
- Human capital: workers' skills and abilities in regards to their contribution to an economy
- Social capital: the value of social networks to society and the human economy
- Natural capital: the resources of an ecosystem that yields a flow of goods and services, also called ecosystem services

Hence, the agriculture landscape is synonymous with natural capital. Natural capital underpins all other types of capital (man-made, human, and social) and is the foundation on which the economy, society, and prosperity is built as it produces flows of ecosystem services that benefit people, business, and society (Natural Capital Coalition, 2015). It is the elements of nature that produce value (directly and indirectly) to people, such as the stock of forests, rivers, land, minerals, and oceans. It includes the living aspects of nature as well as the nonliving aspects such as minerals and energy resources.

3.1.1   Not a new idea

The idea that natural capital is an economic capital can be traced back to Smith's (1952) *Wealth of a Nation* in 1776. In this first economics book, he made a comparison of natural capital of a farm to man-made capital in his statement, "An improved farm may very justly be regarded in the same light as those useful machines which facilitate and abridge labor. An improved farm is equally advantageous and more durable than any of those machines, frequently requiring no other repairs than the most profitable application of the farmer's capital employed in circulation." Regarding the productivity of farming, Smith (1952, p. 157) states that after all their labor, a great part of work always remains to be done by her [nature]. Classical nineteenth century economists such as Ricardo and Faustmann also incorporated the concept of natural resources as capital in economic theory (Fenichel and Abbott, 2014).

In the 1980s Longva was commissioned by Norway's Central Bureau of Statistics to create an economic classification of natural resources. The nonrenewable mineral and energy resources were classified as finite commodities. The renewable biological resources such as fish, forests, and agricultural products were classified as conditionally renewable commodities. As a conditionally renewable resource, their production is dependent on the condition of the natural capital such as the fertility of soil, quality of water, number of reproductive fish stocks, and quality of air.

A third classification was renewable, inflowing resources, associated with the sun and its interaction with the earth's surface and atmosphere. These resources are not traded as commodities and are infinite in supply.

A fourth classification was the environmental resources of air, water, soil, and space. As a rule, environmental resources are not consumed when used in the production process, but as the quality of the resource varies it alters production capacity (Longva, 1981). These are conditionally renewable resources, whose productivity and usefulness, and so their economic values, are dependent on the integrity of the resource.

Repetto (1988) makes a similar, albeit more modern analogy, by stating that as man-made assets (buildings and equipment) age, they become less productive and their worth decreases. To maintain their original or optimum capacity, money is set aside for their replacement in order that the income flowing from their use is sustainable. But the value of natural assets also falls if the assets are depleted. Thus, in the same sense that a machine depreciates, soils depreciate as their fertility is diminished, since they can only produce at higher costs or lower yields. Repetto states natural resource depreciation needs to be treated as asset depreciation in order to put man-made capital and natural capital on the same footing.

Since Longva's (1981) economic classification of natural resources, the field of ecological economics has grown along with the desire to account for and value the *processes* and *functions* of natural capital.

3.1.2   The landscape as a living factory floor

The agriculture landscape has inherent structures, processes, and functions specific to its geography and geology. These inherent attributes are affected by how it is managed. This is analogous to physical or built capital, such as factories. Factories have an inherent structure that supports the processes and functions specific to their design objectives. The outputs and conditions of natural or physical capital are conditionally renewable that are affected by management, which is in turn influenced by how it is accounted for and valued.

3.1.2.1   An automobile factory

For man-made capital, such as an automobile factory, the processes and functions of the factory are easily identified. The price of a car includes the costs of the processes and functions and the structure of the car manufacturing plant. If the factory's equipment degrades, the processes and functions of the factory can no longer produce cars or the factory becomes less efficient in the production of cars. To regain the capacity, financial capital must be reinvested into factory equipment. The accounting strategies for physical capital are well-established business processes as the price of a car includes within it the price of the steel, labor, interest rates, the cost of the factory, computers, the engineering, and the numerous inputs that are needed to construct a car.

To get more cars, people have to demand them (with adequate payment) and in delivering those cars, the processes, functions, and structure of the factory are developed and built. The person who bought the car is also *buying* the processes and functions so that the structure of the manufacturing plant has the capacity to produce cars. The person who wants a car is dependent, indirectly, on those processes and functions of that physical capital structure.

If, in some odd world or culture, cars could not be purchased but were considered public goods, then no one would make cars unless some other component of the car manufacturing plant was valued.

If the processes and functions of the manufacturing plant were valued adequately and there was an efficient method to account for and exchange value for these processes and functions, then attempts to develop the capacity of that manufacturing plant to meet the demands for cars would be valued. This odd scenario to value physical capital as a means to generate the desirable outputs and outcomes of cars is the challenging scenario for valuing natural capital, that is, if one wants to generate the desirable outputs and outcomes of cleaner water and other environmental benefits.

3.1.2.2   A drinking water factory

The purpose of using the *factory floor* as an analogy to the agriculture landscape is to compare how the same components of natural capital, such as the processes, functions, structures, outputs, and outcomes, are valued relative to physical or built capital.

Since 1909, the Seattle Public Utilities has relied on the Cedar River Watershed to purify drinking water for its residents. The landscape, trees, soil, and biology, the bio-geo-chemical processes of the ecosystem, captures, treats, stores, and transports drinking water to Seattle residents. The Seattle Public Utilities views this watershed, in many respects, as their drinking water factory. To mimic these processes, the Seattle Public Utilities would have to build a $200 million filtration plant along with the conveyance system to filter and deliver the city's water supply and budget annual operating and maintenance costs of $3.6 million per year (Cosman et al., 2012).

Despite these comparisons made on scientific principles, the watershed's components related to the delivery of drinking water are not considered economic assets in the same manner as the processes that are constructed from concrete and steel. Cosman et al. (2012) state that if natural capital assets were accounted for, investments in watersheds such as controlling invasive plant or insect species, purchasing additional land that is threatened by development, or helping farmers minimize runoff of animal waste and fertilizer into the watershed would sustain this natural infrastructure resource.

3.2   The conditionally renewable earth factory

To sustain the natural capital outputs of the conditionally renewable resources, the processes, functions, and structures must be properly managed. Since the Earth, as a whole, could be considered a *living* factory, further delineating the ecological structure of the earth factory is needed. At the largest scale, the earth consists of regional biomes with each containing multiple ecosystems. As a living system, ecosystems are built from the bottom up with the processes leading to the structure that support the functions that produce the outputs.

3.2.1 Biomes

Biomes, or "major life zones," are large geographic region of the earth's surface with distinctive plant and animal communities. There are both terrestrial, or land-based biomes, and aquatic biomes. A biome may be spread over a wide geographic area, or as a grouping of many ecosystems that share similar environmental features and plant and animal communities (Caris et al., 1996). Biomes represent a superficial and somewhat arbitrary classification of ecosystems. These are not distinct geographical features such as a mountain range and, therefore, biologists are not unanimous in how they classify biomes or in the number of biomes. Commonly recognized biomes include the following:

- Land-based biomes
  - Tundra (permafrost regions)
  - Taiga (high-latitude forested regions)
  - Temperate deciduous forest
  - Grasslands
  - Deserts
  - Tropical rainforests

- Water-based biomes
  - Marine
  - Estuary
  - Fresh water

Using the factory analogy, the biomes are not the production facilities, but define the regional economy (ecology) and the resources available that support the ecosystems.

3.2.2 Ecosystems

Ecosystems act as the factory unit. They consist of abiotic (nonliving) and biotic (living) components. Abiotic components are sunlight, temperature, precipitation, water, and soil parent material. Biotic components consist of plants, herbivores, carnivores, omnivores, and detritivores (generally those smaller organisms that consume decaying plants). As noted, the entire biosphere of Earth is an ecosystem since all these elements interact. But for analysis and assessment, it is important to adapt a pragmatic view of ecosystem boundaries.

The Millennium Ecosystem Assessment (MEA) developed 10 reporting categories to provide a useful framework for understanding ecosystems relative to human well-being (Alcamo and Bennett, 2003):

- Cultivated: lands dominated by domestic plant species, used for and substantially changed by crop, agroforestry, or aquaculture production
- Dryland: land where plant production is limited by water availability and the dominant uses are large herbivores, including livestock grazing and cultivation
- Forest: lands dominated by trees that are often used for timber, fuel wood, and non-timber forest products
- Marine: ocean with fishing typically a major driver of change
- Coastal: the interface between ocean and land extending into the land to include all areas strongly influenced by proximity to the ocean
- Inland water: permanent water bodies inland from the coastal zone, and areas whose ecology and use are dominated by the permanent or seasonal occurrence of flooded conditions
- Island: lands isolated by water with a high proportion of coast to inland
- Mountain: steep and high lands
- Polar: high-latitude systems frozen for most of the year
- Urban: built environments with a high human density

Defining the boundaries or identifying categories of ecosystems is not an exact science as those definitions are influenced by what the intentions of the delineation are. The MEA categories do not have definite boundaries and often overlap. The MEA uses overlapping categories because this better reflects real-world biological, geophysical, social, and economic interactions, particularly at these relatively large scales.

3.2.2.1 Processes

Within all ecosystems, certain processes occur to sequester, transform, and move energy and materials through the ecosystem. Ecosystem processes include the following (Ruhl et al., 2007):

- Photosynthesis
- Plant nutrient uptake
- Microbial respiration
- Nitrification and denitrification
- Plant transpiration
- Root activity
- Mineral weathering
- Vegetative succession
- Predator–prey interaction
- Decomposition

These processes operate within biological and physical characteristics and constraints. Energy movements are essentially one-way flows that prevent the reuse or recycling of energy. Nutrients, on the other hand, can circulate through different components of an ecosystem and create nutrient cycles and pools. The study of ecology, at its most fundamental level, is quantifying the factors that regulate energy transformation and nutrient cycling within the particular ecosystem defined.

Processes, such as photosynthesis, can be accounted for at multiple scales such as at the level of an individual leaf or within the canopy of a forest. Therefore, defining ecosystem processes must also entail defining the scale at which the process is going to be measured. It should be noted that these fundamental processes exist in all ecosystems whether that is a cornfield in Iowa or the rainforest of Brazil.

A factory analogy would be to consider the processes of man-made capital, such as a factory using electricity, switches, conveyor belts, chains, hydraulics, computers, robots, and other generic processes that other manufacturing plants would use to create their particular products. Each is considered a generic process that exists in virtually all factories, but depending on their particular function, produce a variety of outputs.

3.2.2.2 Functions

Ecosystem functions are the complex physical and biological cycles and interactions that underlie what is observed as the natural world (Brown et al., 2007). The functions of all ecosystems rely on the same biological and chemical processes, but different biome conditions (such as location, soil parent material, climate, etc.) create different functions.

For example, the processes (photosynthesis, transpiration, etc.) in the Iowa cornfield and the rainforest are relatively the same, but the functions become quite different. The corn field has the basic function to produce corn, but can also produce clean water, pheasants, and sequestered carbon. The rainforest has the basic function to create a tree canopy habitat and all the interrelations that make it a rainforest. The same processes within the context of different environments create different functions.

To make a more comparable scenario: a corn field in Iowa that is adjacent to a prairie field has the same processes, location, soil parent material, weather, and climate but they each result in a different function. The ecosystem function is how the processes interact. An analogy would be to consider the functions of two manufacturing plants. One is capable of producing cars and the other dishwashers. Both factories use the same processes (conveyor belts, welding, and hydraulics), but since they are organized differently they have different functions.

3.2.2.3   Structures

Unique to natural systems, the ecosystem functions contribute to the building of the ecosystem's biophysical structure, which in turn supports the functions themselves (Christensen et al., 1996). A crop-management system can create the biophysical characteristics that are advantageous for crop yields. Perhaps more apparent in the rainforest, the structure of the vegetated tree canopy supports the functions of the rainforest. This structure supports the processes and functions of the ecosystem and creates value for humans and their economy. This natural capital system supports the creation and flow of goods and services to humans (Clark, 1995).

3.2.2.4   Outputs and outcomes

The outputs and outcomes that flow from ecosystems are ecosystem goods and ecosystem services. Outputs, or ecosystem goods, are the tangible items such as crops, water, air, and wildlife produced by the agro-ecosystem. While not all ecosystem goods are marketable, they have the characteristics of a physical entity.

Outcomes, or ecosystem services, are the less tangible and most often unmarketable processes such as water purification, nutrient cycling, and pollination. They are the bio-geo-chemical processes of the landscape whose rate and occurrence fluctuates relative to environmental factors and human activities applied to the natural capital. Measuring and accounting for outcomes and outputs is challenging due to the varied nature and interests in the landscape. This is made more complicated as uniform definitions of ecosystem goods and ecosystem services do not exist.

3.3   Varied ecosystem service definitions

Boyd and Banzhaf (2007) noted that a brief survey of definitions reveals multiple, competing meanings of the term *ecosystem service*. This is problematic because environmental accounting systems increasingly are in need of a process and unit strategy to track and measure outputs. The development and acceptance of welfare accounting (i.e., gross domestic product) and environmental performance assessment are hobbled by the lack of standardized ecosystem service units.

Since many stakeholders bring multiple perspectives to landscape sustainability, it is inherent that many definitions and accounting processes emerge. Four prominent strategies include the following:

- All ecosystem outputs and outcomes are categorized under the common heading of *ecosystem services*
- Separately categorizing ecosystem outputs as *ecosystem goods* and outcomes as *ecosystem services*

- Identify outputs based on [geographically based] *service providing units*
- Categorize ecosystem outputs, those ecosystem goods that are directly consumed by humans as *final ecosystem services*

3.3.1 Categorize goods and services as same

The predominant definition today is to refer to both ecosystem goods and ecosystem services as *ecosystem services*. This framework to categorize all ecosystem outputs and outcomes as *ecosystem services* was adopted by the 2003 Millennium Ecosystem Assessment (Alcamo and Bennett, 2003). This strategy is still widely applied, although it is not considered a universally accepted method (Haines-Young and Potschin, 2009). Costanza et al. (1997) used this framework in determining the value of the world's ecosystem services and in an updated 2014 article addressing the changes in the global value of ecosystem services (Costanza et al., 2014). The papers identified 17 types of ecosystem services, which included both outputs, such as crop production, and outcomes, such as water purification.

The MEA (Alcamo and Bennett, 2003) categorized these ecosystem services as

1. Provisional ecosystem services are the consumable products obtained from ecosystems and include food, fiber, fuel, genetics, biochemicals, and fresh water. Provisional services have the characteristic of being a tangible good that is relatively easy measured, valued, and exchanged.
2. Regulating ecosystem services are benefits that are derived from the processes and functions of ecosystems such as water flows, erosion control, water purification, pollination, and biological controls of insects and disease. Regulating services have the characteristic of traditional services that are difficult to measure, value, and exchange.
3. Supporting ecosystem services are those that are necessary for the production of all other ecosystem services, and their impacts on people are either indirect or occur over a long period of time. These include soil formation, primary production, nutrient cycling, oxygen production, seed dispersal, water cycling, and sufficient biodiversity for evolutionary processes.
4. Cultural ecosystem services are nonmaterial benefits that include ecological and cultural connections that create diverse cultures, religious values, knowledge bases, inspiration, aesthetic values, social relationships, sense of place, heritage, and recreation. Perceptions of cultural services are more apt to differ among individuals and communities than how to value provisional services and regulating services.

3.3.2   Categorize goods and services as different

A less prominent definition is based on the traditional economic terms associated with goods and services. Brown et al. (2007) made the distinction that *ecosystem goods* are fundamentally tangible, material, and consumable products such as food, grain, timber, and water generated by the landscape. Like economic goods they are the physical outputs of a process that are produced without interaction with the customers. Goods can be produced to specifications, are relatively stable over time, and their value is associated with the product or unit itself.

Ecosystem services are the less tangible, nonmaterial, and nonconsumable processes such as purifying water, sequestering carbon, and pollination of crops. These ecosystem services occur within and on the landscape and support the production of goods. Like economic services they are intangible processes that cannot be weighed or measured directly and usually require a degree of interaction to have value. Services may vary day-to-day depending on conditions, they can't be stored, and their value is often associated with a location, a good, a person, or site conditions.

She (1997) recognized outputs and outcomes as distinct ecosystem goods and services. She defined ecosystem services as the processes through which natural ecosystems, and the species that make them up, sustain and fulfill human life, as well as support the production of ecosystem goods, such as forage, timber, biomass fuels, natural fiber, and food.

3.3.3   Identify ecosystem services as service providing units

Service providing units (SPUs) uses a geographically based definition and identifies values directly associated with the landscape. SPUs have emerged to ensure that the methods to account for ecosystem services link to and support social and economic values. The SPU suggests that instead of just defining a population or organisms along geographic, demographic, or genetic lines, it can also be specified in terms of the services or benefits it generates at a particular scale (Luck et al., 2009). The use of the word "unit" is an attempt to focus attention on the need to spatially quantify the ecosystem service as it relates to an economic value.

The SPU concept is explained with the example of a population of honeybees. At a given density, the bees may provide all the pollination requirements of an almond grower whose production can be quantified to define the SPU in economic terms. Once an SPU has been defined, attention can be given to how changes in this SPU might affect service production and some measure of value. The definition of an ecological unit is a crucial step to facilitating meaningful economic valuation (Kontogianni et al., 2010) and could support payment programs for specific ecosystem services (Sanchirico and Siikamäki, 2007).

Kontogianni et al. (2010) state advances by ecologists in spatially defining the delivery of ecosystem services can provide a framework for quantifying complex ecosystem processes and resulting services.

3.3.4   Ecosystem services as FEGS and BRIs

The fourth definition described is based on the desire to connect ecosystem good and service value directly to economic indices. A group of scientists at the United States Environmental Protection Agency (USEPA) Office of Research and Development has adopted the concept of Final Ecosystem Goods and Service (FEGS) as a foundation for defining, classifying, and measuring ecosystem services (Landers and Nahlik, 2013) based on ecological endpoints.

Ecological endpoints are defined as the "components of nature, directly enjoyed, consumed, or used to yield human well-being" (Boyd and Banzhaf, 2007). As such, ecological endpoints or final goods are what is counted in GDP: the total value of goods that are consumed. The FEGS system excludes intermediate processes, such as ecosystem services and goods that are used to produce final goods. Using ecological endpoints creates a distinction between environmental features that are directly and indirectly valuable to society. The notion of direct versus indirect goods and services is conventional in the economics of traditional markets (Boyd and Banzhaf, 2007).

As a scenario, Boyd and Banzhaf (2007) state forests that sequester carbon and thus contribute to the reduction of climate-related damages would not be accounted for in the FEGS. They state forests may be a final ecosystem service for other reasons but not for climate-related reasons. In this framework, climate-related damages to natural resources are accounted for due to the effect of climate-related sea-level rise on beach recreation. If sea-level rise damages beaches, and thus recreational benefits, that will be captured in our beach-related ecosystem service measures (e.g., beaches themselves). The fact that forests sequester carbon is certainly important but only in an intermediate sense.

They note that from an economic accounting perspective it does not require the measurement of "all that is ecologically important." Rather, one can economize on measurement by monitoring only the end products of complex ecological processes. It is for this reason that FEGS does not include ecological processes or functions. Relying solely on FEGS would not allow measurement of carbon sequestration and the social cost of carbon as an approach (Olander et al., 2015).

For this reason, beneficial resource indicators (BRIs) are proposed to account for less tangible *nonuse* values that can be difficult to quantify and are often excluded. BRIs are measurable indicators that capture this connection by considering whether there is demand for the service, how much it is used (for use values) or enjoyed (for nonuse values), and whether the particular site provides the access necessary for people to benefit from the service, among other considerations (Olander et al., 2015). BRIs are needed, because while carbon sequestration does not meet the definition of FEGS, it is categorized as a BRI because the research has been done to link carbon sequestration to benefits through the social cost of carbon and the climate.

Hence, from a conceptual perspective, all FEGS are BRIs, but not all BRIs are FEGS.

3.4   Compatible definitions?

The complexity of natural capital outputs and outcomes is complicated by multiple systems of measuring, accounting, valuing, and *defining* ecosystem services. This mix of issues is a cause of the low and diffuse ecosystem service values as well as a reason for high and multiple transaction costs.

The four ecosystem service definitions highlight the disparity among social, academic, ecological, and economical stakeholders in addressing the varied scope and scale of natural capital outputs and outcomes. Each definition is valid in achieving it individual objectives. The MEA's approach considers ecosystem services as the flow of goods to be managed for the well-being of humanity. Brown et al. (2007) and Daily (1997) take a pragmatic approach in describing ecosystem goods and services in the same manner people are accustomed to exchange economic goods and services in day-to-day transactions. Luck et al. (2009) identify ecosystem services within a spatial component that supports associated economic values. Boyd and Banzhaf's FEGS description adopted by a group of EPA scientists identifies human consumption and direct use of ecosystem services as the criteria to account for its economic value within broader economic indicators such as GDP.

Each of the motives of defining ecosystem services is valid and should be considered. In aggregate, they address human use issues, are aligned with typical transactions, and are spatially based and relevant within the broader economic system. Ultimately each of these attributes needs to be incorporated into the definition or the process of accounting for natural capital outputs to allow for measurement, ownership, valuation, and transactions to occur in a *compatible* manner. Compatibility, rather than a uniform or identical accounting and valuation process, seems to be more realistic due to the increasing number of disparate stakeholders participating in defining agriculture sustainability.

chapter four

Disparate stakeholder strategies and values

The second principle issue of the wicked sustainability problem is the increase in the number of stakeholders with diverse values as it relates to natural capital outputs. Today's system as it *is* is the culmination of social and economic influences with a predominant focus on traditional commodities, and a more recent, albeit unorganized, focus on ecosystem services.

4.1 Agriculture's four phases

Timmer (1988) identified four transitional phases of agriculture as follows:

First, there was a pioneering phase in the late nineteenth century in North America, Australia, and New Zealand. A relatively large percentage of the population was farming, and stakeholders consisted of farmers, governments, industry, and localized trade.

Second, the production phase occurred in the early portion of the twentieth century. Organizations providing industry inputs, government support, research for land grant universities, and trade opportunities significantly expanded during this phase.

Third, the productivity phase in the mid- to late-twentieth century focused on improving farming efficiencies. Stakeholder numbers increased due to industrial and conservation progress and the emergent stages of the environmental movement.

The fourth and the current sustainability phase began in the 1980s and was the result of a growing concern for production efficiency, social responsibility, and the environment. This shifted the focus from on-farm production and conservation needs to accounting for off-farm environmental impacts. This seemingly subtle shift in focus resulted in a magnification of natural resource issues of the agriculture landscape and a significant expansion in the scope and scale of organizations interested in agriculture production.

4.2   Stakeholder shift and expansion

Prior to the 1970s, agriculture stakeholders consisted of organizations primarily focused on addressing on-farm production and conservation goals (Sampson et al., 2013). Farmers, ranchers, and organizations directly involved with the research, development, production, and processing of agriculture commodities were the predominant stakeholders. Conservation organizations and federal, state, and local government agencies provided support for soil conservation (Figure 4.1).

Post 1970s, social organizations emerged to address the new focus on sustaining on-farm *and* off-farm resources, food quality, animal health and welfare, and wildlife (Figure 4.2). Scientific studies began to identify the external impacts of industrialization. Urban dwellers adopted policy objectives that focused on health, water quality, environment, recreation, resource conservation, animal rights, food safety, food culture, and other goals. The numbers of interested parties increased, translating into the emergence of political interest groups.

The government responded by creating the U.S. Environmental Protection Agency (EPA) and granting it broad authorities to protect air and water. In the United States, these new viewpoints and value systems brought new and strong positions to farm bill discussions, making reconciliation with traditional agriculture stakeholders difficult in the political process (Sampson et al., 2013).

Figure 4.1 Agricultural policy stakeholders prior to the 1970s in the United States consisted of organizations and individuals relatively close to the production, processing, and resale of commodities. (From Sampson, et al., Solutions from the Land. Report. Edited by Dan Dooley and Kent Schescke, United Nations Foundation, Washington DC, 2013.)

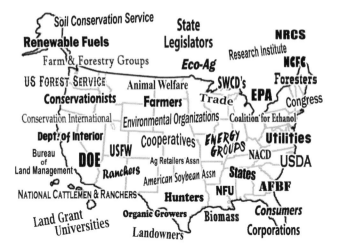

Figure 4.2 Agricultural stakeholders post-1970 in the United States rapidly increased in the number and types of organizations. New organizations formed with support from the sustainability movement to address food, environmental, animal welfare, and corporate supply chains. This was accompanied by the proliferation of livestock, commodity, and bio-energy organizations associated with more global trade and regulatory pressures. (From Sampson, et al., Solutions from the Land. Report. Edited by Dan Dooley and Kent Schescke, United Nations Foundation, Washington DC, 2013.)

Outside the public policy arena, this rapid expansion of social interests often causes conflicting objectives and policies within the organization itself. Food processors and retailers must deal with market forces demanding lower prices, while those same corporations are struggling with defining and delivering sustainability for their customers. The strategy to maximize production in the short term is often at odds with the strategy to maximize natural capital productivity in the long term to resolve water quality, habitat, pollination, and climate issues.

What was once a relatively uniform group of agriculture and conservation organizations focusing on on-farm needs expanded into areas well off of the farm. Policy discussion about working lands once involved those directly associated with the land; now it involves urban dwellers, environmental groups, and a broad mix of nongovernmental organization (NGO) policy organizations, with government taking on a range of new policy agendas that impact land use and land owners.

4.3   Incompatible strategies

As the number of stakeholders increases and the breadth of the issue expands, the challenge to develop the ideal policies or implement the *right* strategies to meet the sustainability goals of each stakeholder becomes overwhelming. Since there are no *right* solutions for wicked problems and since no common platform yet exists to address the breadth of landscape issues, stakeholders often pursue their sustainability goals independently or with some level of cooperation with a limited number of other stakeholders (Zagt and Pasiecznik, 2014).

As more entities begin to value sustainability, they develop varied accounting strategies that increase costs and often add confusion to those managing the landscape. This incompatibility causes the relatively low values associated with ecosystem services to remain diffuse and the transaction costs remain high and multiple.

To describe how this incompatibility emerges, nine case studies representing public utilities, government agencies, industry, NGOs, and corporations are used to explain how their values and strategies differ in achieving the common objective of landscape sustainability.

Sectors and entities include the following:

1. Public Utilities
   a. Seattle Public Utilities and Governmental Accounting Standards Board
   b. Des Moines Water Works
2. Government Agencies
   a. United States Department of Agriculture (USDA) Farm Bill Conservation Title
   b. Chesapeake Bay TMDL
   c. Minnesota Department of Agriculture—Agriculture Water Quality Certainty Program (AWQCP)
3. Industry
   a. United Suppliers
4. NGO and Corporations
   a. The Sustainability Consortium
   b. Field to Market®
   c. EPRI Ohio Basin Water Quality Trading

4.3.1   Public utility sector

Public utilities provide energy, communication, sanitary, and transportation services as well as drinking water and water treatment services and often rely on landscapes outside their jurisdiction to meet the community needs. The American Public Works Association (APWA), an organization representing public work professionals, defines sustainability in public works to mean delivering services in a manner that ensures an appropriate balance between the environment and the communities that future generations will reside in.

Accordingly, the *EPA's Planning for Sustainability: A Handbook for Water and Wastewater Utilities* calls on drinking water and wastewater utilities to ensure that water infrastructure investments are cost-effective over their lifecycle, resource efficient, and consistent with other relevant community goals (EPA, 2012).

While the term *natural capital* was not used in either document, the term and concept are emerging in public utility efforts as included in APWA's Center for Sustainability objectives, including restoring the environment and advocating for community sustainability (APWA, 2013). The following two public utility issues relate to accounting for and managing natural capital relative to the landscape providing a drinking water supply.

4.3.1.1   Seattle Public Utilities and the GASB

The Seattle Public Utilities is a government-owned utility providing typical public services including drinking water for Seattle residents with the mission to deliver clean, affordable water to its residents. The Seattle Public Utilities purchased the forested Cedar River Watershed, which has been successfully delivering clean water to Seattle residents since 1909. As the utility's greatest asset, the Seattle Public Utilities would like to value the Cedar River Watershed and its ecosystem functions within the context of GASB's accounting standards. The GASB is a private, nonprofit independent organization that establishes and improves standards of accounting and financial reporting for U.S. state and local governments. It has a broad mission to create an accounting framework that results in useful information to guide the public, issuers, and auditors of financial reports (GASB, 2015).

The Seattle Public Utilities and nine other utilities seek a formalized natural capital accounting method from the GASB, which does not recognize natural capital in the same manner as physical capital (i.e., a filtration plant). Because of this omission, the Seattle Public Utilities cannot secure loans against the natural capital values or make improvements to it under accepted accounting standards. The Seattle Public Utilities claims the current GASB accounting rules do not provide an accurate or meaningful picture of the water utility's assets.

Cosman et al. (2012) state the value of the ecosystem services delivered by the watershed can be estimated by considering how the functions of the watershed would need to be replaced to provide clean water by other means. This estimation can take into account the construction and maintenance costs of filtration plants, plus the costs of obtaining water from another source, such as desalinization or groundwater pumping. To replace the processes of the Cedar River Watershed, the Seattle Public Utilities would have to build a $200 million filtration plant along with the conveyance system to filter and deliver the city's water supply and to budget the annual operating and maintenance costs of $3.6 million per year. This $200 million filtration plant along with pipes, vehicles, buildings, roads, computers, copy machines, and fences would count as an asset on their books. The value of the forested watershed filtering and transporting water and meeting the same need does not count.

This Seattle Public Utilities scenario revealed three important points as it relates to natural capital (Cosman et al., 2012):

1. Natural capital tends to provide benefits over a very long period (centuries or longer), whereas man-made capital provides benefits in the near term (years to decades).
2. Natural capital appreciates in value over a long period due, in part, to increased scarcity and functionality, whereas built capital depreciates relatively rapidly.
3. Investments in natural capital with the goal of sustainability can be far better investments over the long term than investments with shorter, but less sustainable benefits.

The disparity between the Seattle Public Utilities and the GASB can be further described with a theoretical example of a water treatment facility. When an infrastructure asset such as a filtration plant is constructed, its capital asset value will be listed as the cost (e.g., $10 million). If the plant has a 20-year life and is depreciated evenly at $0.5 million/yr, then the asset value at the end of the first year will be $9.5 million. At the end of year 10, the value is $5 million. If the plant is upgraded, these capital expenses would increase the asset value. So, a $2 million expense to improve the plant in year 10 would then place the asset value at $7 million; the capital expense ($10 million) minus depreciation ($5 million), plus improvement ($2 million).

When this accounting strategy is applied to natural capital, three issues emerge. First, the cost of acquiring or improving green infrastructure may be low. For example, the cost of a tree seedling for watershed restoration is minimal. Second, natural capital investments such as a tree seedling literally grow and appreciate in function and value. Traditional accountant methods do not recognize "negative depreciation." In this case, the asset value of a tree would never exceed its very low establishment cost. And, third, maintenance may not be considered as capital expenses. Consequently, using this standard methodology, "cost, less depreciation," does not result in a representative value of natural capital, as envisioned by these public water utilities (Hartel, 2003).

Cosman et al. (2012) concluded that many sustainability issues have their roots in uncontrolled depletion and damage of natural capital as a result of an outdated and inadequate approach to economics. If natural capital were valued similarly as physical capital, utilities would quickly recognize the need to protect, repair, and enhance the function of their watersheds. These capital enhancing activities, for example, could consist of controlling invasive plant or insect species, purchasing additional land that is threatened by pollution-generating development, establishing vegetation, or helping farmers minimize runoff of animal waste and fertilizer into the watershed.

4.3.1.2   Des Moines Water Works and county drainage boards

In January 2015, Des Moines Water Works (DMWW) notified the three Iowa counties of Buena Vista, Calhoun, and Sac upstream of a primary drinking water supply for 500,000 residents that it intends to file a federal lawsuit under the Clean Water Act (CWA) to stem nitrate pollution in the Raccoon River (Stowe, 2014).

Similar to the Seattle Public Utilities, the DMWW depends on a watershed to provide a reasonably clean source of drinking water. In contrast to the Seattle Public Utilities, the DMWW does not own the watershed and relies on others to manage the landscape to provide the water quality needed for cost-effective treatment.

Since 1919, the DMWW has operated its public utility to process and provide drinking water from the Raccoon and Des Moines River watershed to the people in Des Moines and surrounding communities. Water from the Raccoon and Des Moines Rivers are economic drivers that benefit Central Iowa by providing abundant sources of water to citizens, business, and industry. The rivers are a source of recreation for people who canoe, kayak, and fish on the water trails (DMWW, 2015). The land in the watershed is owned by residents, governments, and businesses of the three counties.

The 10 drainage systems within the three counties are managed by the county board of supervisors of those counties and are the parties the DMWW intends to sue. The foundation of the lawsuit is based on the conveyance of nitrates through tile drainage pipes and ditches associated with ground water. While the CWA exempts agriculture stormwater drainage, the DMWW claims the spikes in nitrate concentration is the result of groundwater conveyance and not stormwater. Since the source of the nitrates is not stormwater, the suit claims the drainage systems are not exempt from the CWA and therefore, a National Pollution Discharge Elimination System (NPDES) permit is required.

The lawsuit claims the primary relief sought is to bring the drainage systems in compliance with permit requirements, although it states relief from costs related to nitrate removal upgrades.

Unlike other dischargers of pollution, the agriculture producers associated with the drainage systems value nitrates as a production input. In other words, from a crop production efficiency and profit perspective, agriculture producers do not want nitrates leaving their fields. Since nitrogen chemically changes into several forms depending on the site conditions and ecosystem processes, managing nitrogen levels in soil and water is very challenging in the field, in the water, and at the treatment plant. This pending lawsuit will likely have an impact on what level of responsibility land managers have in relation to the natural capital outputs generated (Clayworth, 2014).

4.3.2 Government agency sector

Government agencies in the United States are extensively involved in issues related to agriculture landscape management at the local, state, and federal levels. The USDA, EPA, and the Department of Interior are federal agencies with policies and programs influencing commodity production, water quality, and habitat management to name a few. Various local and state agencies are delegated federal agency authority or provide cooperative services to administer federal goals.

The following examples describe three strategies: (1) the nationwide USDA Farm Bill Conservation Title, (2) the regional Chesapeake Bay Program, and (3) a state-level (MN) regulatory certainty program. Despite all being government agencies, different strategies are used to meet similar water quality objectives.

4.3.2.1 USDA federal farm policy

Approximately every five years the U.S. Congress passes federal farm policy, often referred to as the Farm Bill. The 2014 Farm Bill consists of 12 titles with titles I, II, and XI (commodity, conservation, and crop insurance, respectively) being the most notable titles affecting landscape management decisions (Johnson and Monke, 2014).

Title II, the Conservation Title, is the title predominantly focused on sustaining agriculture landscapes. It includes the Environmental Quality Incentives Program (EQIP) that pays for individual conservation practices on working, productive farmlands; the Conservation Stewardship Program (CSP) that pays for outcomes and individual practices on working, productive farmlands; and the Conservation Reserve Program (CRP), a land rental program that takes land out of commodity production and pays a rental rate based on an Environmental Benefits Index using a calculation containing soil, carbon, vegetation, and wildlife criteria and data. The Agricultural Conservation Easement Program (ACEP) is an easement purchase associated with restoring wetlands and preserving farmlands. Each of these conservation programs relies on different strategies to value and account for landscape sustainability objectives using practice-based and index-based accounting methods.

Title I, the commodity title, provides payments based on a decline in either price or revenue (market price times crop yield). Title XI, the crop insurance title, provides subsidies in the form of partial payments to cover insurance costs for producers who purchase a policy to protect against losses in yield, crop revenue, or whole farm revenue.

The farm bill titles and programs often have conflicting conservation and production objectives, in that one title may provide incentives that oppose the other title's objectives (USDA, 2014a).

4.3.2.2   EPA's Chesapeake Bay Protection and Restoration Order

The Chesapeake Bay Protection and Restoration Executive Order–Executive Order 13508, signed by President Obama on May 12, 2009, called for the development of a system of accountability for tracking and reporting conservation. The Executive Order describes the full accounting of conservation practices applied to the land as a necessary data input for improving the quality of information and ensuring that the practices are properly credited in the Bay model (White House Press Secretary, 2009).

The seven watershed jurisdictions are required to account for best management practices (BMPs) and manage new or increased loadings of nitrogen, phosphorus, and sediment. A BMP Verification Review Panel was established to provide advice, feedback, and recommendations to the Chesapeake Bay Program partnership as it develops its verification program. By 2018, the seven jurisdictions are required to account for BMPs implemented in the sectors of agriculture, forestry, stormwater, wastewater, wetlands, and streams by using a practice-based accounting system.

4.3.2.3   Agriculture Water Quality Certainty Program

The Minnesota Agriculture Water Quality Certainty Program (AWQCP) is a program born out of a state–federal partnership that includes the Minnesota Department of Agriculture (MDA), the USDA Natural Resource Conservation Service (NRCS), and the EPA. A Memorandum of Understanding was signed on January 17, 2012, by Minnesota Governor Mark Dayton, U.S. Agriculture Secretary Tom Vilsack, and EPA Administrator Lisa Jackson.

The AWQCP certifies farms based on the installation of conservation practices and a water quality index (WQI) score of 8.5 on a scale of 1–10. The WQI score was developed by NRCS and was modified to address natural resource needs specific to Minnesota. The WQI is calculated by using several landscape factors related to soil, slope, cropping systems, weather, and conservation practices (MDA, 2015).

4.3.3   Agriculture industry sector

A variety of agriculture industries are involved in supplying inputs and services for the production of commodities that are processed into foods and consumable goods. Increased pressure from food retailers and consumer groups has motivated industries to develop strategies to account for or improve the sustainability of their product supply chain and manage environmental risks.

4.3.3.1   United Suppliers' SUSTAIN

United Suppliers was established in 1963 as a cooperative by 30 Iowa retailers joining forces to manufacture feed for livestock. Today, United

Suppliers is a member-owned wholesaler that provides agricultural products and services to about 700 grower cooperatives and retailers covering 45 million acres of farmland in the United States.

In July 2014, United Suppliers initiated the SUSTAIN program to promote the 4R program consisting of the critical concepts of right rate, timing, source, and placement of fertilizer to ensure higher nutrient use efficiencies and less nutrients lost to the water and air. United Suppliers partnered with the Environmental Defense Fund to develop and implement a fertilizer efficiency program to meet the supply chain demand for Walmart and others (Toot, 2014).

By offering specialized sustainability services to growers, United Suppliers expects to reduce fertilizer losses on 10 million acres of cropland by 2020. Other organizations associated with SUSTAIN are Walmart, General Mills, Smithfield, and Coca-Cola Company (Environmental Defense Fund [EDF], 2014). The SUSTAIN program uses performance indicators that aggregate producer data to account for sustainability outcomes sought after by suppliers to Walmart and other retailers.

4.3.4   NGOs and the corporate sector

In the early years of the environmental movement, nonprofit NGOs and for-profit corporations usually had conflicting values associated with landscape sustainability. More recently NGOs and corporations are developing collaborative relationships yielding benefits for both the corporate and NGO participants and the general welfare of the populations of concern.

These relationships provide corporations with access to different resources, competencies, and capabilities than those that are otherwise available within their organizations (Yaziji and Doh, 2009) and it aids to build their images and reputations (Burgos, 2012). Similarly, these connections may provide NGOs access to financial and nonfinancial resources and expertise from those corporations, with whom they collaborate, including managerial and technical skills, marketing leverage, and other capabilities (Yaziji and Doh, 2009).

These relationships have advanced agriculture sustainability at the farm and landscape scale by identifying the common objectives of managing natural capital for sustainability goals. The following examples describe three methods to account for and value natural capital outputs. The Sustainability Consortium is developing a process from a product perspective and seeks data upstream from its suppliers and their suppliers. Field to Market is developing an outcome-based process from the field downstream to the product. And the third example is the Electric Power Research Institute and their water quality trading program using a practice-based accounting system.

4.3.4.1   The Sustainability Consortium

The Sustainability Consortium is a multistakeholder, collaborative approach with members representing corporations, NGOs, academic institutions, trade organizations, and expert representatives of civil society. It is jointly administered by Arizona State University and the University of Arkansas with additional operations at Wageningen University in The Netherlands and Nanjing University in China. Its purpose is to create consumer product sustainability by working across sectors and geographies at multiple scales and scopes.

It attempts to accomplish this by using a standardized reporting framework for the communication of sustainability-related information throughout the product value chain.

The reporting system is composed of a portfolio of three components: a product category dossier, a category sustainability profile, and key performance indicators (KPIs). These products provide a traceable, consistent, practical, and science-based method for understanding the key environmental and social issues within the life cycle of a product category (The Sustainability Consortium [TSC], 2015a,b).

Supplier performance is based on questionnaires in which suppliers respond to certain practices applied and outcomes generated by supplier and producer activities. By using this tool, companies can improve the quality of decision making about product sustainability to effectively manage the sustainability of upstream supplies and suppliers, while communicating product sustainability downstream to consumers.

4.3.4.2   Field to Market

Field to Market is a diverse alliance working across the agricultural supply chain to make continuous improvements in productivity, environmental quality, and human well-being. The group provides collaborative leadership that is engaged in industry-wide dialogue, grounded in science, and open to the full range of technology choices.

This alliance consists of a diverse group of producers, agribusinesses, food, beverage, restaurant, and retail companies, conservation organizations, universities, and public sector partners seeking to create sustainable outcomes for agriculture. With this collaborative effort, Field to Market is helping to define, measure, and advance the sustainability of food, fiber, and fuel production using indices as outcome-based approaches.

The Field to Market metrics estimate field-scale performance using data such as land use, soil erosion, soil carbon, irrigation water use, water quality, energy use, and greenhouse gas emissions, and then compares individual producer's results against benchmarks calculated from publicly available data.

This data can also be repurposed to create shared value throughout the supply chain. For example, corn growers in Nebraska working with

Bunge, Kellogg Company, Extension, and NRCS can understand their sustainability trends and their individual sustainability performance, as well as opportunities for improving practices (Field to Market, 2014).

4.3.4.3   EPRI water quality trading

The Electric Power Research Institute, Inc. (EPRI) is an independent, non-profit organization whose work spans nearly every area of electricity generation, delivery and use, management, and environmental responsibility. Its Ohio River Basin Interstate Water Quality Trading Project is a collaborative effort to improve water quality in the Ohio River Basin through implementing an interstate water quality trading program.

Tradable credits are generated by reductions in nutrient runoff achieved by implementing additional BMPs beyond the current land management or baseline management. In simplest terms, the agricultural baseline sets the bar that must be achieved by a farm before that farm can generate credits. Credits are generated by installing BMPs and tracked with a practice-based accounting system.

The need for a water quality trading program is based on the anticipation of new or more stringent numeric water quality criteria, total maximum daily loads (TMDLs), and/or water quality-based NPDES permit limits. At full scale, this project could span portions of eight states and create a market for thousands of point sources to purchase nutrient credits, and involve approximately 230,000 farmers in selling credits (EPRI, 2012).

4.4   Following Carlson's Law?

The proliferation of sustainability efforts across sectors and by several organizations may seem both promising and troubling. Many stakeholders that need to become aware of the value of natural capital and take action are doing so, but their independent efforts cause additional confusion and inefficiencies as to what each are attempting to accomplish. This result is aligned with *Carlson's Law*, a term attributed to Curtis Carlson, former CEO of SRI International, to describe an effect caused by now widely available data, networks, and communications. Carlson stated that in a world where so many people now have access to cheap tools of innovation, innovation that happens from the top down tends to be orderly but dumb and innovation from the bottom up tends to be chaotic but smart (Friedman, 2011). Carlson used this law to describe the balance between autocracy and democracy in an organization. The term *dumb* in this case of agriculture sustainability efforts refers to a *not smart* system where accounting and measurement methods are not aligned or integrated with the activities associated with producing the desired outcomes. In other words, the checklists and other assurance processes are not integrated with the production plans of many farmers.

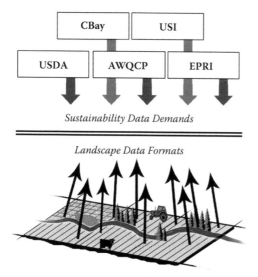

Figure 4.3 Agriculture sustainability efforts are prone to experience Carlson's Law effects when orderly, top-down, dumb processes are not coordinated with bottom-up chaotic and smart processes. Precision agriculture data is *smart*, as it is integrated and supports the management activities of the land managers, but they emerge from many perspectives and each data stream is uncoordinated. The accounting systems of the corporate and government sustainability efforts are orderly, but often do not reflect the needs of the land managers.

Because agriculture production and sustainability affects so many organizations and individuals, the effect of Carlson's Law is amplified. Rather than a single hierarchy-type organization with their own top-down processes, there are multiple organizations with hierarchy, market, or network organizational structures with each having their own version of a top-down process. Figure 4.3 attempts to illustrate this disconnection between top-down and bottom-up strategies.

The resulting bottom-up chaos does not interface effectively with the multiple rigid and disparate top-down processes.

4.4.1   Top-down [dis]orderly

Despite the orderly trait of top-down innovation, the sheer number of organizations interested in agriculture sustainability creates disorder from the top down. To paraphrase and apply Carlson's Law to multiple top-down efforts essentially creates *chaotic* top-down dumb systems. It should be noted that despite the crude terminology, these top-down orderly innovative efforts are a key to the longer term success of sustaining agricultural landscapes. It is not intended to imply the pioneering innovations of these organizations are in any way detrimental to progressing sustainability, but that they must at some junction integrate and be able to interface with the many bottom-up innovations. This recognition becomes more apparent by the number of entities that may begin to demand sustainability outcomes and data from utilities, governments, and corporations:

- There are approximately 161,000 public water systems in the United States with many relying on surface waters similar to the Seattle Public Utility and DMWW watersheds. Pending issues such as the development of GASB natural capital accounting standards or the legal proceeding of the Des Moines Water Works case against the Iowa country drainage systems could provide new legal and financial avenues for utilities to create more accounting strategies (EPA, 2009).
- The Chesapeake Bay is one of 43,000 waters determined to be impaired by the EPA. Impaired waters are required to develop a TMDL plan and the Chesapeake Bay TMDL is one of 68,000 TMDLs in the United States addressing various impairments in those waters (EPA, 2015). Because there is not a "one size fits all" approach to restoring and protecting water resources, states are able to develop tailored strategies in the context of their water quality goals. Many of the TMDLs could develop unique accounting strategies. In response to this growing TMDL issue, several states are beginning to implement water quality certainty programs similar to the Minnesota AWQCP. Presumably, each state could develop an accounting and valuation program independently from other states.
- Corporations and businesses total into the hundreds of thousands with each potentially having unique supply chain needs and sources. Specific product categories have varied environmental impact resulting in unique questionnaires and assessments to determine product sustainability. Corporations having different perspective, values, and value systems may view and define sustainability in very different manners.
- Government agencies are prone to similar challenges as corporations. Differing units of government have different objectives and perspectives and presumably would define sustainability by using unique methods. A commonly cited example of government agencies using different and conflicting objectives is the commodity and conservation titles within the USDA farm bill. Similar scenarios are repeated across local, state, and regional governments and vary by how an agency addresses specific landscape objectives.

Within a single organization, Carlson's Law may cause inefficiencies, but the result of it applied to a social, wicked problem such as agricultural landscape sustainability reveals an emerging unwieldy scenario of multiple top-down protocols. The complications resulting from this disorderly process are magnified as hundreds or thousands of individual agriculture operations and associated business attempt to address these emerging sustainability demands.

4.4.2   Bottom-up chaos

The other effect of Carlson's Law is the numerous responses from the bottom up, often from the practitioners of the organization. The practitioners respond with innovation to address the new top-down demands, but do so in a way that is aligned with their needs and activities.

In response to sustainability demands, the agriculture community has developed data collection, environmental assessments, and accounting systems at the field, farm, watershed, and supply chain scales. A Google search of agriculture sustainability reveals numerous business software and accounting systems for agriculture sustainability. Referred to as *precision agriculture*, these systems are "smart" in that they address issues relative to the individual's needs and activities. Because bottom-up approaches often includes market-based systems, competition among the stakeholders prevents a sharing of information and strategies. The result is many customized land management accounting systems with the tendency to create confusion and chaos.

This scenario is being manifesting in accounting for and managing nutrients for crop and forage production. One system may be developed to account for the 4R Program, a production and environmental-based program to adopt the right rate, right source, right timing, and right place for nutrient use. Another system may be based on Adapt-N, a program to account for the movement and loss of nitrogen fertilizer, while others may focus on nutrients and soil loss. Each has the capacity to contribute to the overall sustainability objectives, but also contribute to the disparity of systems. The end result is numerous accounting methods for a variety of production and natural resources which when compiled create some level of chaos. It should be noted that these forces of chaos are recognized, leading agriculture industry efforts such as the Open Ag Data Alliance (OADA), a platform to integrate precision agriculture data (OADA, 2014).

4.5   Disparity rooted in culture and governance

The increasing numbers of stakeholders and their strategies pose significant issues related to transaction costs and accounting methods. This issue of disparate systems is, in part, rooted in organizational culture and governance.

Governance, that is how things get done within an organization, usually has far more influence on how an organization interacts and achieves objectives than most individuals within and external to the organization realize. This cultural dimension of governance has long been neglected (Meuleman, 2013), yet agricultural landscape sustainability and the science of sustainability demand that governance becomes an integral part of devising wicked solutions. Due to these types of dynamics, Jessop (2002) states the topic of governance is clearly a notion whose time has come.

chapter five

Conflicting governance styles

The final component of the wicked agriculture sustainability problem is the conflicting governance styles of stakeholder organizations. Conflicts among governance styles arise due to an organization's expectations on decision making, delegation of responsibility and power, verification of performance (Meuleman, 2008), and methods of acquiring knowledge (Nickerson and Zenger, 2004). Conflicts between agricultural sustainability organizations arise, not only due to differences in sustainability objectives, but due to pursuing *common* objectives under different governance styles.

5.1 Organizations

Organization is a generic term that describes a variety of different aggregations of people and structures with a final goal (Avril and Zumello, 2013). There are a variety of organizations, including corporations, governments, nongovernmental organizations, not-for-profit corporations, partnerships, cooperatives, universities, and various types of political organizations. Organizational types, like governance styles, have evolved over time to adapt to societal challenges.

5.1.1 Organizational evolution

Early organizations, such as fiefdoms and gang-like groups, relied on a hierarchical structure with division of labor controlled by violence. The mafia, drug cartels, and gangs using this hierarchical structure reside at the fringe of society and in the illegal trades today (Laloux, 2014).

Organizations have and continue to evolve as they address more complex issues. The gang-like structures evolved into conformist-based hierarchies to maintain control as they scaled up in size. To solve more complex issues, an incentive-based hierarchy emerged, common to corporations. And today, pluralistic hierarchies and nonhierarchical network structures have emerged to address multidisciplinary and more complex social issues (Laloux, 2014).

5.1.1.1   Conformist hierarchy

A state-centric model or a conformist-based hierarchy evolved to address the unstable nature of violence-based gang-like structures and to effectively scale up in size. Conformist hierarchies are represented by bureaucracies such as government, public schools, churches, and the military. Promotion within the organization is based on seniority with stability valued above all (Scharmer and Kaufer, 2013; Laloux, 2014).

5.1.1.2   Incentivized hierarchy

A free-market model or an incentive-based hierarchy evolved from the conformist hierarchy to solve issues of competition and promote innovation. Like the conformist hierarchy, it uses a formal command-and-control to determine the organizational objective, but allows freedom on how the objectives are met. Promotion is based on solutions and profits rather than solely seniority. These types of organizations are represented by corporations and others relying on innovation, accountability, and merit (Scharmer and Kaufer, 2013; Laloux, 2014).

5.1.1.3   Pluralistic hierarchy

A social-market model or a pluralist hierarchy is a top-down management organization with a cultural-driven focus. It empowers employees with decision-making abilities. Management is value-driven and negotiated from organized interest groups and multiple stakeholder perspectives. These types of organizations are represented by corporations such as Southwest Airlines and Ben & Jerry's Ice Cream and emerging B-corporations (Scharmer and Kaufer 2013; Laloux, 2014).

5.1.1.4   Network structure

A cocreative ecosystem model (Scharmer and Kaufer, 2013) or a nonhierarchical evolutionary network organization (Laloux, 2014) is purpose-driven. It relies on self-management and the entity, as a whole, determines growth and change strategies. While relatively rare, Laloux (2014) cited a dozen companies, large and small, across sectors to describe this emerging type of organization structure.

5.1.1.5   Organizational structure–governance styles connections

Organization types and governance styles are closely linked. The three typical governance styles are hierarchy, market, and network. These governance styles are aligned with the hierarchical and networked structure of organizations. Where they differ is that governance is not a structure, but a process on how objectives are accomplished. As a process, governance styles are more fluid than an organizational structure. Organizations often use a mix of governance styles to acquire information and accomplish internal and external objectives.

5.2   Governance

Governance refers to processes of governing, whether undertaken by a government, market, or network (Bevir, 2009). Broadly defined, governance is the whole of interactions that government, the private sector, and civil society participate in to solve societal problems or create opportunities (Meuleman, 2008). Governance is a broader term than government, and in its widest sense it refers to the various ways through which social life is coordinated. Government may be one of the institutions involved in governance, but it is possible to have governance without government (Rhodes, 1996).

Harlan Cleveland is credited for first using the word *governance* as an alternative to the phrase public administration as he stated the people want less government and more governance (Frederickson, 2004), implying governance occurs in the broader social arena.

Governance is the sum of the many ways individuals and institutions, public and private, manage their common affairs. It is a continuing process through which conflicting or diverse interests may be accommodated and cooperative action taken. It includes formal institutions, as well as informal arrangements that people and institutions either have agreed to or perceive to be in their interest (Roe, 2013; Ostrom, 2007). Governance emerges from formal and informal organizational structures, such as families, teams, churches, businesses, and government entities. The structure of an organization or the culture of a societal sector becomes the framework from which governance emerges.

5.2.1   Governance cultures

Governance of an organization is like a social culture and is viewed as belief systems. Davis (1984) defines organization culture as the pattern of shared beliefs and values that give members of an institution meaning, and provide them with the rules for behavior in their organization. These close cultural connections create difficulties in any attempts to change an organization's governance style. "Culture eats strategy for breakfast," a phrase originated by Peter Drucker and made famous by Mark Fields, President at Ford, sums up the challenge of changing organizational governance (Rick, 2014).

Meuleman (2008) credits Aaron Wildavsky as one of the main scholars who focused on reintroducing the importance of culture to political sciences. Wildavsky described how the three socially active ways of life (hierarchism, individualism, and egalitarianism) are aligned with the three typical governance styles: hierarchy, market, and network. Each governance style also includes behavior rules: regulations and control instruments in hierarchy, competition in markets, and trust in networks (Meuleman, 2008). Within this context, Thompson (1991) describes governance styles as how organizations "get things done." This culture connection is why organizations sometimes fiercely defend their governance style and why some organizations consider their governance styles as a panacea for all problems (Meuleman, 2008). These governance styles create competition among organizations, often in a hostile way, but on the other hand require one another, and therefore continue to coexist. This governance paradox of conflict and cooperation is a product of recent governance evolution.

5.2.2   Governance styles

Governance styles can be defined as the processes of decision making and implementation, including the manner in which the organizations involved relate to each other (Kersbergen, 2004). Governance styles, like organizational structure, have evolved from hierarchy to market to network governance styles as issues have increasingly become more complex (Laloux, 2014). Figure 5.1 illustrates an evolutionary timeline of emerging governance styles in Western European nations during the last century (Meuleman, 2008). Hierarchy governance has its origins in early society while market governance emerged in the 1980s and network governance styles developed relatively recently in the 1990s. The timeline illustrates another interesting phenomenon; never before in human history have organizations working side-by-side on common issues had such diverse governance styles (Laloux, 2014).

Hierarchy, the oldest governance style, was a term used historically to describe the structure of the Church and is derived from two Greek words: *hieros*, which means sacred, and *archein*, which means rule or order. The term *hierarchy* now refers to a form of organization in which people and functions are organized in a well-defined, multitiered, vertical structure with

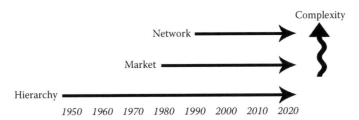

Figure 5.1 Governance has evolved from predominantly top-down hierarchy styles to more innovative market governance and to more inclusive network governance. Hierarchy governance was dominant into the later part of the twentieth century. Market governance was adopted by governments and institutions as social issues became more complex. Network governance styles emerged in the 1990s as society became increasingly interconnected. (From Meuleman, L., *Public Management and the Metagovernance of Hierarchies, Networks and Markets: The Feasibility of Designing and Managing Governance Style Combinations*, Physica-Verlag, Heidelberg, Germany, 2008, p. 43.)

different functional tasks performed at each level in the structure (Bevir, 2009). Hierarchical governance is often considered an *old* style of governance, whereas network and market governance have been referred to as the *newer* modes of governance (Börzel et al., 2005). Market governance emerged in the 1980s from a movement called *New Public Management* governance which was initiated to address government inefficiencies by bringing management concepts from private business into the public sector (Kersbergen, 2004).

Network governance, emerging in the following decade, is a nonhierarchical process that incorporates a horizontal structure encompassing various organizations. A network shares information and often operates without defined hierarchical leadership (Bevir, 2009).

Despite the most recent governance styles being more complex, Laloux (2014) cautions against the tendency to assume the more recently evolved types of organizations and governance styles are better than the previous styles. The more recent organizations are able to deal with more complex situations, but not all situations require or are best handled at higher levels of organizational complexity.

5.2.2.1 Hierarchy governance

The hierarchical mode of governance was broadly adopted in Europe to replace *arbitrary authoritarianism and nepotism* in government and to provide a means for standardizing government tasks (Meuleman, 2008). It was adopted in the private sector in the United States in the late nineteenth century with the advent of mass production (Avril and Zumello, 2013). From the early years of industrialization, businesses increased the scale of their operations by developing hierarchical organizations that vertically integrated different stages of production. Government, university, and nonprofit institutions followed suit and organized themselves in much the same way (Morrison, 2012).

The hierarchical structure was designed and best suited to manage complicated processes such as automobile assembly where production could be broken down into a series of steps. Hierarchical corporations were organized so that centralized control could direct the entire production process. Although effective for managing large numbers of workers, such structures lacked agility (Avril and Zumello, 2013).

The key attributes of hierarchy governance are as follows (Meuleman, 2008, p. 22):

1. Carefully defined division of tasks
2. Impersonal authority vested in rules that govern official business
3. Recruitment of employees based upon proven or at least potential competence
4. Seniority and merit-based promotion
5. Subordinate-superior authority structure

The basic assumption which generates bureaucratic hierarchical structures is that each member is restricted to a single specialized task with each level controlled by the next higher level (Herbst, 1976).

5.2.2.1.1   Advantages

Hierarchical governance fosters stability, predictability, rule of law, and clear lines of command (Meuleman, 2014). This form of governance became the basic pattern for public administration in many Western democracies as it provided a way for standardizing government tasks (Herbst, 1976), reporting relationships, and assigning responsibilities. These defined boundaries deliver rationality, clarity, and stability, particularly as job descriptions become narrower and focused, and employee skills become more specialized (Morrison, 2012).

5.2.2.1.2   Limitations

A government-centered, hierarchical governance system is often too rigid and can inhibit communication within the organization. These communication breakdowns can create a sense of fragmentation and tension which undercut coordination and productivity. Excessive specialization can build up formal procedures that slow internal operations. The organization becomes less responsive to changes taking place in its environment due to this inflexibility (Morrison, 2012).

Modern multiactor, multisector, and multilevel problems are too "fuzzy" for inflexible hierarchical governance to manage (Castells, 1996). Traditional top-down hierarchical governance is not capable of adapting as contemporary organizational environments have become more complex and interconnected (Avril and Zumello, 2013). This accelerating change allows less time for passing orders and reports up and down a hierarchy. People on the front line need to be able to respond without delay within an agreed framework of goals, vision, and values (Rosell, 2000).

Hierarchical governance began to lose some of its attraction in the 1980s when the market governance movement or *New Public Management* became the focus of both public administration scholars and practitioners (Meuleman, 2008) and so organizations with top-down hierarchical governance styles sought new strategies (Avril and Zumello, 2013).

5.2.2.2   Market governance

The term *market* in market governance refers to the types of mechanisms and thinking associated with the private sector, such as competition and customer service. It should not to be confused with the economic market or the governance of the private sector market (Meuleman, 2008). Market governance emerged in the 1980s from *New Public Management*, which was initiated to address government inefficiencies by bringing management concepts from private business into the public sector (Kersbergen, 2004).

Market-based governance includes delegating traditional governmental functions to the private sector, and applying market-style management approaches and mechanisms of accountability to government functions (Donahue and Nye, 2002). In other words, government agencies are contracting out some tasks to the private sector and adopting the business model of the private sector to accomplish the tasks they retain.

Market governance promotes the modernization of public institutions with new forms of management aimed at creating efficient administrations. The key attributes of market governance are the following (Fábián, 2010; Alonso et al., 2013):

1. A service-and-result centered organizational structure
2. A value-producing management process adding new value at each step
3. Efficient data collection and processing
4. Effective political and administrative management
5. Competitive elements to achieving objectives

The adoption of key words such as market, client, competition, and especially corporate management symbolizes a movement away from strict hierarchical administrative management (Fábián, 2010). It is an attempt to focus on improving efficiency by creating businesslike service delivery with results-based, value-driven outputs and the managerial freedom to accomplish these objectives (Christensen and Lægreid, 2011).

5.2.2.2.1 *Advantages*

Market governance seeks to increase competition in the delivery of public service through organizational restructuring to become customer oriented (Alonso et al., 2013). Market governance promotes competition between service providers and empowers citizens by pushing control out of the bureaucracy, into the community. An outcome-based focus is driven by goals and is favored over an input focus or strictly on rules and regulations. It attempts to redefine clients as customers and offer choices to seek out solutions, and to prevent problems before they emerge, rather than simply offering services afterward (Klijn and Koppenjan, 2000).

In its ideal-typical form, it stimulates the formation of hybrid organizations (mixtures of public sector and private sector organizations), and emphasizes the management competencies of staff, instead of policy-making competences. It stimulates benchmarking and contract management and advocates for outputs (Meuleman, 2008).

5.2.2.2.2 *Limitations*

Market governance may be challenged to address the different circumstances of private and public operations. The field of operation of the private sector is market and proprietorship, while for the public sector, it is democracy and the rule of law (Fábián, 2010). This may lead market governance to be more about efficiency than effectiveness (Vries, 2010).

Democratic accountability may be diminished, not only because elected officials under market governance lose their top-down authority over public bureaucracies and managers, but also because it is difficult to maintain and increase the bottom-up control of all officials, including those employed on contracts as well as elected officials (Kersbergen, 2004).

Often, the private sector is dominated by single-person decisions, whereas the majority of decisions are made collectively in the public sector. Privatization and delegation to independent agencies give citizens fewer chances to control such agencies through voice (Kersbergen, 2004). Outcomes may not be evenly distributed if equity considerations are of less concern than under traditional bureaucracy (Fábián, 2010).

Market governance may have the tendency to create scenarios where solutions are short term and nondurable, and so responsibilities fall back to the traditional hierarchical structure. The adoption of market governance strategies may also suggest that the private sector is superior to the public sector and create low morale in public administration (Meuleman, 2008).

With the market governance shortcomings and the limitations of traditional hierarchical organizations becoming clearer in a world of rapid change, there is a growing emphasis on developing informal relationships, open networks, and temporary partnerships (Rosell, 2000).

5.2.2.3 Network governance

Network governance provides a third alternative between top-down planning of hierarchies and the profit focus of the market. It is seen as a response to the failures of markets and hierarchical coordination and is enabled by recent societal and technological developments (Provan and Kenis, 2007). From a functional point of view, the aim of network governance is to create a synergy between different competences and sources of knowledge in order to deal with complex and interlinked problems.

Networked governance, as a decentralized, integrative form of problem solving, is promising because it allows actors outside of government to contribute their unique resources to the generation of creative, collaborative, and complex solutions (Huppé et al., 2012). Broadly defined, network governance consists of relationships between interdependent actors from all levels of government, and from various political and societal groups (Bevir, 2009; Meuleman, 2008).

The popular view of networks as a flat, horizontal mode of organization is very one-sided. Networks are only "flat" in comparison with hierarchies. Networks may also have centers and central modes of steering and governance (Dijk and Winters-van Beek, 2009) and consist of interconnected nodes of different sizes that are linked by relationships of varying degrees and strengths.

Five key attributes (Klijn and Koppenjan, 2000) of network governance are as follows:

1. Actors are mutually dependent, which leads to sustainable relations between them.
2. Rules are formed during the course of interactions.
3. Policy processes are complex and not entirely predictable because of the variety of actors, perceptions, and strategies.
4. Policy is the result of complex interactions between actors who participate.
5. Conflict management and risk reduction are managed through network cooperation processes.

Organizations using network governance reject the command-and-control strategies associated with hierarchic bureaucracies and rely on negotiation and trust. They see the role of the state as a facilitator or enabler and help foster partnerships with and among public, voluntary, and private sector groups. Citizens are not merely voters or consumers of public services but are active participants within such groups and policy networks (Bevir, 2009). Unlike hierarchy and market governance, networked organizations lack a predetermined structure, which often creates a low level of understanding how networks work. However, analysis of network governance has grown to the point where it is now widely recognized to have considerable validity over and above that of the traditional hierarchy (Roe, 2013).

Network governance emerged as society became interconnected and the policy-making arena became sectoralized, coinciding with a broadening and more decentralized scope of state policy making (Roe, 2013). In this environment, three forms of network governance configurations developed (Provan and Kenis, 2007):

1. Lead: a highly centralized and brokered form with a single participant within the network providing administration and facilitation of network activities.
2. Administrative: a highly centralized and brokered form with an external organization providing administration and facilitation of network activities.
3. Participant: a highly decentralized form governed by the network members with separate and unique governance entities. Governance can be accomplished formally, through regular meetings, or more informally through the ongoing, but typically uncoordinated efforts of those with a stake in the outcomes.

Network governance is appropriate for particular problems, but it is not a solution for everything. It is able to manage complexity, enhance flexibility, support innovation, and personalize relationships. Network governance requires a particular management approach that is only beginning to be discovered (Dijk and Winters-van Beek, 2009).

5.2.2.3.1 *Advantages*

Network governance is seen as a strategy to increase coordination and coherence in public policy (Christensen and Lægreid, 2011) and offers advantages for learning and innovation in an ever-changing environment (Meuleman, 2008). The advantages of network governance in both public and private sectors are considerable, including enhanced learning, more efficient use of resources, increased capacity to plan for and address complex problems, greater competitiveness, and better services for clients and customers (Provan and Kenis, 2007).

Advocates of network dialogue and deliberation argue that network governance facilitates social learning. In their view, public problems are not technical issues to be resolved by experts. Rather they are questions about how a community wants to act or govern itself. Dialogue and deliberation better enable citizens and administrations to resolve these questions (Bevir, 2009).

Networks provide a kind of dynamism and flexibility that hierarchies cannot, and yet foster cooperation and stable relationships in a way that market governance cannot. Advocates say these advantages are especially relevant in today's complex and interconnected world (Bevir, 2009). The most important characteristic of using networks as a mode of governance is the combination of horizontal and vertical control and coordination (Dijk and Winters-van Beek, 2009).

5.2.2.3.2 *Limitations*

Governing by networks is complex and the efficacy of networked governance is thus in constant flux. A slight change in the network can generate sufficient shifts and disrupt the possibility of an effective collaborative process. The process of networked governance itself introduces an additional component of complexity. This complexity, if unmanageable, can undermine the problem-solving process (Huppé et al., 2012).

Roe's (2013) identified five criticisms of network governance:

1. Lack of a theoretical foundation: It is not based on a solid body of knowledge and its concepts are unclear.
2. Lack of explanatory power: Specific outcomes may not be known until the process is completed.

3. Neglect of the role of power: It does not address the inherent conflicts associated with power and control.
4. Lack of evaluation criteria: It does not offer a clear framework for evaluating outcomes.
5. Discounts governmental role: It often views government agencies to be the same as other organizations and neglects their role as a guardian of public interest.

Under conditions of insufficient social capital, engaging in networked governance may erode the fabric of trust and collaboration, subjecting participants to conflict, destructive opportunism, and power struggles, further entrenching a confrontational (us vs. them) mentality (Huppé et al., 2012).

Kamarck (2007) cautions against broadly using network governance since it works best on policy problems that require flexibility, personalization, and innovation, and other problems could better be solved by different governance styles. Another weakness inherent to networks is that people with a higher than average number of "links" with others play a more important role in networks and can establish a kind of hierarchy in a network. These hubs provide communication, but if such hubs are removed, networks may break down into isolated pieces (Meuleman, 2008).

5.2.3   Governance conflicts

Jessop (2002) states that each governance style has its own distinctive forms of failure, and the combination of the three ideal-typical governance styles lead to conflicts, competition, and unsatisfactory outcomes. This conflict is the result of each governance style having its own logic as it relates to relationships, decision making, and compliance (Roe, 2013).

1. Relationships: Hierarchies are centrally controlled and based on dependency and subordination. Market governance consists of a virtually infinite number of independent actors. Network participants have many interdependent relationships between actors.
2. Decision making: Hierarchies are centralized and top-down. Markets are directed by profits and innovation. Networks are characterized by ongoing negotiation and achieving collective solutions.
3. Compliance: Hierarchies achieve compliance through rules and laws. Markets rely on the fear of economic loss and networks rely on trust.

Three problems emerge as the logic and application of these governance styles converges. First, their strategies can undermine each other. Second, each of them has typical failures or even unreasonableness. Third, they all have an attractive logic that leads each of them to believe their governance style is best. The latter relates to the close ties governance styles have to organizational culture (Meuleman, 2013). These cultural *clashes* occur as organizations use different strategies to obtain knowledge and achieve objectives. For example, as a means to determine standards, hierarchy seeks routine, market governance seeks price signals, and networks seek relationships (Dijk and Winters-van Beek, 2009).

Even with common objectives, conflicts arise from the sets of values held by these governance styles. Hierarchy governance values are based on the expectation that there should be a *subordinate* to the hierarchy and it relies on regulations and control instruments to meet goals. Market-based governance values a *customer* perspective and relies on competition and innovation to achieve results. Network-based governance seeks *partners* and *cocreators* and relies on trust to achieve outcomes (Meuleman, 2008).

These conflicts are not based on *what* should be done, but on *how* things should be done. The source of these conflicts is the three organizational cultures whose conflicting governance styles create a *trilemma*.

5.2.3.1   Governance trilemma

Meuleman (2008) describes the governance trilemma as the need to solving three interconnected dilemmas: among hierarchy and network, hierarchy and market, and network and market. Figure 5.2 illustrates the interactions resulting in each of these dilemmas and the tendency for organizations to pull toward their specific apex. Each governance style *seeks* its innate approach (i.e., network consensus, hierarchy rules, and market competition), resulting in incompatible strategies.

Shell (2005) describes a trilemma as a trade-off between three competing forces that results in a "two-wins, one-loss" scenario. Shell uses the term *force*, rather than *governance*, at the triangle apexes, but their terms align with the three governance styles: coercion and regulation are aligned with the hierarchical style; market incentives and efficiency are aligned with the market style; and social cohesion and community are aligned with the network style.

Shell uses an example to show how the three forces cannot be completely appeased. It states that while societies often aspire to efficiency (market), social cohesion and justice (network), and security (hierarchy), they are not mutually inclusive as one cannot be at the same time freer, more conformant to one's group, and more coerced. The best one can expect is to capture the most plausible trade-offs between these diverse, complex objectives, namely in which forces combine to achieve more of two objectives.

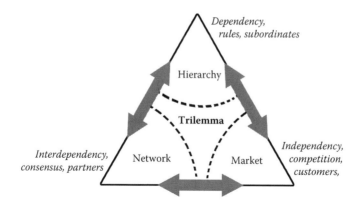

Figure 5.2 The governance trilemma occurs as the attributes from each governance style pressure the other two. Hierarchy seeks top-down and subordinate control, market seeks competition and customers, and network governance seeks consensus and partners. Conflicts among and between governance styles occur as competing characteristics interact when dealing with relationships, decision making, and compliance. (From Meuleman, L., *Public Management and the Metagovernance of Hierarchies, Networks and Markets: The Feasibility of Designing and Managing Governance Style Combinations*, Physica-Verlag, Heidelberg, Germany, 2008.)

Meuleman (2008) states that even with the most plausible trade-off, third-party issues need to be addressed as they may endanger the outcome favoring the other two. An example is the often observed trade-off between the relatively new modes of governance (network and market) in environmental policy, which is a threat to the idea that the environment should also be protected by legislation through hierarchy governance.

Jessop (2005) sees another aspect of the governance trilemma and points to situations when agents are faced with choices such that new governance solutions may undermine key conditions of their existence and/or their capacities to realize some overall interest.

5.3 Governance conflicts in the system as it is

Conflicts arise in the agricultural system as it *is* (Figure 2.3 in Chapter 2) as many sectors and multiple organizations with various governance styles interact while exerting their influence on agriculture landscape management decisions. In many scenarios and within the context of the case studies reviewed, it appears that little or no recognition is paid to an organization's governance style or if the governance styles are compatible. Governance of the organizations is so embedded in history and culture that governance actors are generally not aware of the governance style they operate within. Without this knowledge or context, transforming the system as it *is* to a system as it *ought to be* does not seem feasible.

Patterson et al. (2015) state that there is a need for a broad framework to understand and analyze important dimensions of governance in relation to transformations toward sustainability. This is important for exploring the role of governance, and allowing cross-case analysis and comparison to build a higher-level theory over time. Achieving sustainability objectives inherently requires transdisciplinary attempts (Stock and Burton, 2011) and so, in devising a wicked solution, case studies representing potential transitions away from the system as it *is* toward a shared governance system as it *ought to be* are examined.

section two

Devising a wicked solution

chapter six

Devising a wicked solution

The first step in devising a wicked solution is recognizing the problem as a wicked one (Commonwealth of Australia, 2007). This sounds obvious, but many complex problems are viewed as complicated, solvable problems; and attempts are made to address wicked problems by recasting them as tame ones. By treating a wicked problem as a tame problem, energy and resources are misdirected, resulting in solutions that are not only ineffective, but also can actually create more difficulty (Nelson and Stolterman, 2003).

Rittel and Webber (1973) state that finding the solution is intimately tied to the problem, to the extent that finding the problem is the same as finding the solution; the problem cannot be defined until the solution can be found. And the information needed to understand the solution and the problem depends on one's idea for solving it.

Since wicked problems are part of the society that generates them, any resolution brings with it a call for changes in that society. These changes include different forms of governance, new methods of research, and decision making based on that research (Brown, 2010). Instituting these changes requires the application of transdisciplinary science. This process requires experts from multiple disciplines, together with policy makers, stakeholders, and representatives of various public entities, to produce practice-oriented knowledge for addressing complex societal problems (Buizer et al., 2011)

6.1   A transdisciplinary approach

There is an ever-increasing call from society and the scientific community for transdisciplinary approaches to tackle fundamental societal challenges, especially those related to sustainability (Lang, 2012). Interest in transdisciplinary research has burgeoned in the last 10 years (Gray, 2008). Transdisciplinary research differs from interdisciplinary research (which focuses on integration across disciplines) as its goal is to transcend disciplines by focusing on real-world complex problems through collaboration between academic and nonacademic stakeholders (Patterson et al., 2015). A transdisciplinary approach is essentially a participatory, interdisciplinary process whose success is founded by building joint visions of the issue of concern, finding a common language, and discussing the trade-offs that result from particular choices, and above all through collaborative learning (Jager, 2008).

Lang (2012) defines transdisciplinarity as a reflexive, integrative, method-driven scientific principle aiming at the solution by using knowledge from various scientific and societal bodies. This definition highlights that transdisciplinary research requires (1) a focus on addressing real-world problems and a (2) collaboration with different disciplines and actors from outside academia, and looks (3) to create knowledge that is solution-oriented, socially robust, and transferable to both scientific and societal practice (Lang, 2012; Patterson et al., 2015). Transdisciplinary research is necessary because understanding, analyzing, and contributing to transformations toward sustainability cuts across academic disciplines, policy domains, and societal sectors (Patterson et al., 2015).

6.1.1   Acquiring system knowledge

Transforming an agriculture production system requires fundamental knowledge concerning natural and social factors and processes, and their connections. In their report, *Research on Sustainability and Global Change— Visions in Science Policy,* Swiss researchers (Conference of the Swiss Scientific Academies, 1997) state that three types of knowledge are required:

1. Systems knowledge of structures and processes, and variability (the observed system as it *is*)
2. Target knowledge of the desired system (the system as it *ought to be*)
3. Transformational knowledge on how to make the transition from the current to the target situation

Transdisciplinary research is much more than interdisciplinary data gathering; research in this sense is also rooted in imagination and intuition (Jackson, 2008). A transdisciplinary approach explores the research questions at their intersections to allow stakeholders to invent new science together to reframe their conceptual frameworks. Such reframing requires the suspension of current assumptions in which the previously unthinkable can become reality (Gray, 2008). To resolve a wicked problem, one requires a new mental model that has not been imagined before (Brown, 2010).

6.2   Imagination as a wicked solution strategy

Imagination is essential for the construction of mental models or representations of reality that people use to understand complex phenomena (Jackson, 2008). Imagination is associated with creativity, insight, vision, and originality and is also related to memory, perception, and invention. All of these are necessary in addressing the uncertainty associated with wicked problems (Brown, 2010). Coming to terms with wicked problems not only requires knowledge and intelligence, but imagination as well (Fulton, 2011).

Using examples in the formalized sciences, Brown (2010) states how the imaginations of Linnaeus, Aristotle, and Darwin all contributed to the understanding of life forms. These leaps of creativity were essential for them and others to imagine the complexities of biological processes and are essential in solving transdisciplinary, wicked problems. One of the greatest critical thinkers of modern times, Albert Einstein, stated imagination is more important than knowledge. Referred to as *thought experiments*, Einstein imagined himself chasing a light beam, which was credited to his eventual comprehension of the theory of relativity (Norton, 2015). These thought experiments are necessary for new concepts as knowledge is limited to all we now know and understand, while imagination embraces the entire world, and all there ever will be to know and understand.

Brown (2010) states that despite these common experiences, scientists are accustomed to thinking imagination is the enemy of scientific research and discounts it as the source and creative spark for scientific inquiry. In seeking solutions for wicked problems, researchers and decision makers are brought together, each with radically different understandings of the issue. To make linkages between them, they must use their imagination to place themselves in others' shoes to identify problems and solutions.

CogNexus Institute founder Jeff Conklin sees imagination as the key to addressing research *fragmentation*. Fragmentation refers to the individual perspectives, understandings, and intentions of the stakeholders, all of whom are convinced that their version of the problem is correct. The inability to imagine the situation of other stakeholders is a significant challenge to resolving wicked problems (Conklin, 2006).

6.3   Identifying wicked solution sources

If imagination can spark a solution idea for a wicked problem, it can only provide a plausible starting point and from there one must experiment, learn, and adapt. Transdisciplinary collaboration, by definition, carries an effort beyond traditional disciplinary study and into the real world. Without this strategy, many efforts get struck with *analysis paralysis*, a condition that prevents any further action until all is known about the task at hand (Govindarajan and Gupta, 2001). At this point, execution is more important when innovation is at the heart of a strategy, since innovation involves treading into uncertain waters and pursuing entirely new models. The greater the uncertainty of a problem and the solution, the lesser will be the value of a well-thought-out strategy.

Six case studies developing new models for addressing agricultural landscape sustainability were chosen to provide transdisciplinary insights and a source for a wicked solution. None of the case studies were initiated with the stated goal of using a transdisciplinary approach to solve the wicked problem of sustaining landscapes. The stakeholders that initiated each of the projects did so to address a specific need relative to their organizational values and objectives.

The six projects chosen for the agricultural landscape sustainability case studies share a unique connection geographically, politically, and chronologically. Geographically, all occurred within the borders of the state of MInnesota in the United States. Politically, they each had a broad spectrum of stakeholders and collectively involved diverse stakeholders such as agriculture producers, local, state, and federal agencies, nonprofit organizations, and industry representation. In total, stakeholders included dozens of public, private, and nongovernmental entities and hundreds of practitioners. Chronologically, the projects occurred from 2001 to the present day with minimal overlap and gaps in their delivery.

6.4   Six pilot project case studies

Individually, the six case studies sought new ways to account for and value natural capital, to address the varied sustainability concerns of stakeholders, and to develop new intersector relationships to meet their objectives. In some cases, they reassessed their traditional ways of working and solving problems. This reassessment of protocol is often necessary as wicked problems are often unsolvable with existing governance structures, skill bases, and organizational capacity (Commonwealth of Australia, 2007).

Each project had different motivations and intentions for addressing agricultural landscape sustainability. The research paradox of using these case studies is that while they have a rather limited geographical scope, somewhat contrary to the broad vision one often desires in transdisciplinary research, their relatively close proximity in time and space allowed knowledge transfer to occur with relative ease. This proximity and the chronological nature of the projects (2001–present) aided in compiling and devising a wicked solution, as the six cases, in aggregate, resembled a longer-term transdisciplinary approach with adaptive management characteristics.

The six case studies are listed in chronological order; organization and project name are included:

1. 2001–2009; Minnesota Milk Producers Association (MMPA): Environmental Quality Assurance Program (EQA)
2. 2006–2008; Minnesota Project: Conservation Planning for Agronomic Service Centers
3. 2007–2008; Minnesota Pollution Control Agency: A Conservation Bridge for Agriculture Professionals

4. 2010–2011; Minnesota Department of Agriculture: Livestock Environmental Quality Assurance Program
5. 2012–present; Minnesota Department of Agriculture: Ag Water Quality Certainty Program
6. 2013–2014; Chisago Soil and Water Conservation District (SWCD): Sunrise Watershed AgEQA Project

Each case study approached this common issue of agriculture sustainability using unique strategies developed from multiple stakeholder perspectives and governance styles. To analyze the case studies for transdisciplinary insights, five questions from Brown's (2010) open transdisciplinary inquiry perspective were used:

1. Focus: What is the problem being addressed?
2. Context: What are the views of the stakeholders?
3. Sources of evidence: What are the stakeholders' knowledge bases?
4. Synthesis framework: How was evidence and knowledge compiled and analyzed?
5. Collective learning: What were the findings (partial, uncertain, and open-ended)?

The purpose of this inquiry is to view the case studies with a transdisciplinary lens and to use the new knowledge to create a rich picture of the agriculture landscape system as it *ought to be* (Williams and Van 't Hof, 2014). A rich picture describing the system as it *ought to be* contains the target knowledge consisting of the new configuration of stakeholders, components, value flows, and outputs assumed to be the new basis for agriculture sustainability.

6.4.1   MMPA's EQA

The MMPA initiated an environmental quality assurance program (MMEQA) in 2001 with a focus on developing greater statewide consistency in conservation technical support for dairy farmers and to certify farms when they met state and industry-developed environmental quality regulations (LCCMR, 2001).

The context of the project included the range of agricultural production and sustainability perspectives from the agriculture, conservation, government, and environmental sectors. This was the knowledge base and evidence used to create the on-farm environmental assessment. As the lead stakeholder, the MMPA had final approval of the resource assessment template, training program, and assurance process. More than 300 dairy farms were assessed by MMPA-trained technicians using a practice-based assessment consisting of five categories: water quality, odor and air quality, soil quality and nutrient management, habitat quality and diversity, and community image.

The findings were that the assessment process was a good learning tool for dairy farmers, but the data collected was cumbersome (not scalable) and did not serve the accounting and assurance purposes of any of the other stakeholders. The MMPA program did create a broader technical base for dairy farmers through the training program. A conclusion was that there is a growing interest for some type of agriculture sustainability assurance process, but it needed to be more comprehensive, inclusive, and streamlined.

6.4.2　Minnesota Project's conservation innovation

The Minnesota Project, a nonprofit organization, received a 2006 United States Department of Agriculture (USDA) Conservation Innovation Grant (MP-CIG) to implement a project (*Conservation Planning with Agribusiness Centers*) to train agronomic professionals to become Natural Resources Conservation Service (NRCS)–certified conservation planners. The focus of the project was to increase private sector technical assistance capacity in response to the passage of the Conservation Security Program (CSP): an outcome-based incentive program of the 2004 USDA Farm Bill. This project had a similar objective to the MMEQA of increasing technical assistance for agriculture producers, but for different stakeholders.

The context of the MP-CIG varied from the MMEQA, as several of the primary stakeholders (state government agencies, agribusinesses, and conservation organizations) questioned the need or demand of the training program. Politically, the government conservationists viewed the role of conservation planning to reside with government agencies, not private sector agronomists. They also recognized the private sector was profit-centered and did not envision farmers paying for a service that was provided free from the USDA and partners. Several agriculture groups thought this may expand conservation requirements for farmers. In contrast to their concerns, some private sector agronomists viewed this as a progressive approach to address emerging environmental needs or as a new business opportunity.

Despite these perceived obstacles, 28 agronomists attended the 9-day course over a 9-month period. The trainings were presented by professionals in government, agriculture, academia, and business. They consisted of classroom and field exercises on the importance and management of agriculture landscape resources.

The findings of the effort included the realization that only a small percentage of agronomic professionals would be able to become efficient conservation planners, but the majority was capable of becoming efficient assessors of on-farm natural resources. Due to their existing role, agronomists had trusting relationships with the farmers, knowledge of the landscape, and insights into the production system of the farm operation. Since the majority of the information needed to conduct resource assessments is collected by agronomists, a natural resource assessment became a cost-effective endeavor for the agronomist, in contrast to being a very costly endeavor for government conservation staff that often have to develop a relationship with farmers from the beginning.

From a process perspective, it was found that an on-farm assessment based on land management indices was a more efficient method than the practice-based accounting of the MMEQA. It was also an effective method to educate agronomists and farmers on government program objectives and create a convenient "dashboard" to provide public and private resource professionals a means to quickly identify on-farm conservation concerns.

This case study identified the index-based resource assessment as a pivotal component in integrating production resource and natural resource management at the farm level. The indices were developed within the scientific communities of government and academia and applied within the technical community of agronomy for the benefits of agriculture practitioners and government conservation planners. The index-based platform provided support for a transdisciplinary approach to agriculture landscape sustainability. It provided the platform for academic and nonacademic participants to engage in sustainability outcomes (NRCS CIG, 2005).

6.4.3 Minnesota Pollution Control Agency's conservation bridge

The 2007 project, *A Conservation Bridge for Agriculture Professionals* (CBAP), was developed to address the perceived need for more private sector conservation advisors and to identify costs. With the emergence of corporate sustainability efforts, carbon and water quality trading schemes, pending water quality regulations, and expanding government conservation incentives, it was proposed that agriculture producers need direct access to a larger field of natural resource professionals.

The focus of the project was to identify the costs and benefits of agriculture professionals conducting index-based on-farm resource assessments. The project objectives were to employ the trained conservation professionals from the MP-CIG project and use a similar index-based assessment template. The context and stakeholder views were narrower in scope compared to the MP-CIG, but the stakeholder types were similar. Agriculture advisors were paid US$400 per assessment to account for the soil, water, and habitat resources of the farm operation.

A survey on the ease and ability of completing these natural-resource assessments for the 30 farm resource assessments was conducted by the agronomists and their farmer clients.

The farmers surveyed stated the following:

- Understood the concept of using management indices to rate their farm operation
- Liked the concept of measuring the management of their farm operation for themselves and others, but several were concerned about who would get the information, how it would be used, and other privacy issues
- Generally agreed with the "measurement" results
- Could see value in using this process as developing a related conservation plan document
- Felt that this type of resource-assessment process could be integrated with their production plans

The resource assessors surveyed stated the following:

- Understood the concept of using management indices to rate farm operations
- Liked the concept and felt it could be very beneficial for many of the resource issues farmers are facing at the local to national level
- Generally agreed with the results but felt that some of the indices need to be further refined
- Thought the process would be helpful for them to develop conservation plans and felt they would be more proficient after doing several of these
- Felt that this process could be incorporated into a farmer's production plans

This case study reinforced the findings from the MP-CIG project that an index-based resource assessment was a cost-effective means to generate new resource management knowledge and to integrate the objectives of farmers, agronomists, and government agents. It also aided the agriculture community in understanding the objectives of government conservation programs (MPCA EA, 2010).

6.4.4 *Minnesota Department of Agriculture's EQA*

In 2009, the Minnesota Legislature passed Chapter 172, Article 2, Section 2 to fund the Livestock Environmental Quality Assurance Program (LEQA) for the purpose of providing resource management analysis, assistance to create an implementation plan, and assurance the plan was completed.

The focus was to create the means for livestock farmers to assure their farm operations met the state's water quality objectives relative to watershed total maximum daily load (TMDL) plans. TMDLs are watershed goals stating how much of a particular pollutant is permitted to be in a waterbody. Similar to the MMEQA, the program development involved a broad base of agriculture, conservation, and environmental stakeholder perspectives. It also provided training for public and private agriculture and conservation professionals to conduct on-farm assessments and assurance. In contrast to the MMEQA, the narrative assessment was converted, in part, to an index-based assessment. This conversion was based on the findings of the previous case studies.

The findings of this project included the creation of *landscape intelligence*, the result of using index scores associated with a parcel of land, creating an *assessment unit* that could be aggregated at multiple scales. The potential of landscape intelligence was revealed when a regional cheese processor and an electric utility inquired about the use of the LEQA model and its indices to account for sustainability objectives. Each was seeking a method to account for sustainability of their products of cheese and electricity. In discussion, it was discovered that each entity could express their own sustainability criteria independently yet use similar indices. In theory, it would allow multiple entities to procure sustainability data in a more cost-effective means than if each entity adopted noncompatible assessment processes.

The synergies of this process extended beyond just the financial costs of data collection and analysis, but also included a means for others to participate in the processes of landscape sustainability with relative ease. This cooperative process was recognized as a shared governance model with the August 2011 report entitled, *Using a Shared Governance Model to Certify Minnesota's Clean Water Legacy Farms* (Gieseke, 2011).

6.4.5 *EPA–USDA–MN Ag Water Quality Certification Program*

In January 2012, the Environmental Protection Agency, USDA, and Minnesota's Governor signed an agreement to develop the Minnesota Agricultural Water Quality Certification Program (AWQCP). The focus of the project was to develop a regulatory assurance process and a 10-year contract option for agriculture producers to ensure water quality sustainability.

The context and knowledge base of the stakeholders consisted of a state–federal partnership that included the USDA, EPA, Minnesota Department of Agriculture (MDAg), Minnesota Pollution Control Agency (MPCA), Minnesota Board of Water and Soil Resources, and Minnesota Department of Natural Resources. The Minnesota Legislature passed a 2014 statute further defining the purpose and goals (MN Leg, 2014).

Program input was provided by agriculture, conservation, environmental, government, and industry representatives to pilot the project in four watersheds. State and local governments collect and manage the database and make adjustments to the index parameters, calculations, and other components as necessary. Local conservation district staffs run calculations for the water quality index (WQI) using soil data, topography, cropping system, fertility, and pest management.

The preliminary findings were that similarly to the LEQA, the WQI scores associated with a management unit (field) can be compiled and organized in various methods to create landscape intelligence. GNP Company, a chicken processor, is participating in the program and is using the landscape intelligence as a means to account for sustainability outcomes for their brand (Redlin et al., 2015).

6.4.6   *Chisago SWCD's AgEQA*™

The Sunrise Watershed Project (AgEQA) was jointly initiated by the Chisago SWCD and the MPCA with a focus on increasing conservation technical support capacity for producers in the watershed and on creating a method to account for water quality improvements in the Sunrise Watershed.

The context of the program development included input from the Chisago SWCD, local agronomists, NRCS staff, the MPCA, and Ag Resource Strategies, LLC, a private business. An AgEQA portfolio was created using indices related to water, soil, plants, nutrients, and wildlife within the farm management units of farmscape, farmstead, livestock facilities, and fields. The indices were scaled from 0 to 100 with 100 being the best management for a particular resource.

The portfolio, consisting of natural resource index values, was viewed as a representation of the farm's ecological assets. The value of the assets could be improved by adjusting landscape management strategies. The index-based portfolio could be used as a common platform for the farmer, agronomist, conservation agent, and the MPCA to discuss and value these ecological assets.

This local project was presented by Kimble (2014), UN Foundation Vice President, at a Global Land Tool Network meeting at The Hague in June 2014. She noted, "What the experts in Europe and elsewhere had not focused on is the multiple layers of regulatory policy in the U.S.—and few understood the distinction between EPA's objectives and those of USDA. [The way this project] approached the information makes it possible start a land evaluation at a more rudimentary level and build up. This approach would work well in Brazil where there are some differential policies between states like Minas Gerais and the federal government. It was a good complement to an Australian presentation on dairy farm management—which highlighted some similar challenges."

This unique scenario, where a relatively small and local project was introduced to an audience with a global perspective, revealed the commonality of issues faced among agriculture sustainability stakeholders and practitioners.

6.5 A [compiled] transdisciplinary approach

The close geographical, political, and temporal relationships among these six case studies provided the context of a transdisciplinary approach. Each was focused on addressing real-world problems and collaborating with different sectors, disciplines, and actors from outside academia. Because these six projects occurred chronologically over a decade, new knowledge from one project could be readily incorporated into the next project. It created an adaptive management atmosphere that was solution-oriented and allowed for the social acceptance of new ideas over time.

6.5.1 Retelling the transdisciplinary story

Compiling the six projects and retelling them as an aggregated decade-long effort provides insights and generates transformational knowledge. Transformational knowledge enables practitioners to evaluate different problem-solving strategies; to achieve the competence to foster, implement, and monitor progress; and to adapt and change behavioral attitudes (Hadorn et al., 2006). Transformational knowledge is the basis for the decisions on how to transform the system as it *is* to as it *ought to be*.

This transdisciplinary story began with a milk producer representative seeking new ways to increase technical assistance to help dairy farmers meet environmental goals and account for natural resource management. The MMEQA on-farm assessment was beneficial in helping producers understand their environmental issues, but the narrative format was a bit too challenging to aggregate information or to convince other entities to formally recognize the certification.

The MP-CIG proposal was also in response to the lack of conservation technical assistance for producers but from the perspective of a new federal conservation program. The assessment process incorporated environmental indices, rather than a narrative assessment format to streamline data collection and management. It was concluded that conservation planning was too challenging to incorporate into the agronomist workload, but it was recognized that a stand-alone index-based resource assessment process would be an efficient method to integrate the expertise of agriculture professionals into the conservation delivery system.

With this new knowledge, the CBAP project analyzed the costs and willingness of agronomic professionals to conduct on-farm index-based assessments. Both were deemed favorable as long as the agronomist was currently working with the producer. The LEQA program adopted the index-based assessment process on a field basis and created landscape intelligence. This data could be customized to address sustainability objectives for a variety of stakeholders such as state government, a cheese processing plant, and an electric utility. This common platform revealed a shared governance model that allowed engagement by disparate stakeholders with common, but not identical goals.

The AWQCP applied the index-based process specifically to meet state water quality objectives. Similarly to the LEQA, but in a more formal manner, corporate sustainability stakeholders recognized it as a usable valuation platform to account for their sustainability claims as it related to water quality.

The Sunrise AgEQA project compiled multiple indices to address the broad spectrum of natural resources and calibrated the indices on a 0–100 scale to act as a farm portfolio dashboard. This portfolio approach was recognized at the international level as a potential common platform to address agriculture sustainability at the local and global scale.

6.5.2   Transformational knowledge

Transformational knowledge is the knowledge required to understand how to transition a system from as it *is*—the system knowledge—to as it *ought to be*—the target knowledge. It refers to the knowhow, the tacit knowledge that can be derived from target and system knowledge. System knowledge refers to the current state of a system, its underlying drivers, and its ability to change. Target knowledge determines the design of a plausible system that may meet the desired outcomes.

To achieve sustainability outcomes, future environmental and sustainability research needs to place significantly greater emphasis on target and transformation knowledge as well as system knowledge in the human and social sciences (CASS, 1997). Imagining and retelling the six case studies as one transdisciplinary story revealed several key evolutionary points. The story began as an effort to address the lack of technical assistance and to create a means to account for agriculture sustainability. The story ended with three key findings: (1) an index-based training module to increase the technical assistance available from the private sector, (2) a dashboard accounting system and portfolio, and (3) scalable landscape intelligence in the form of index values that could be exchanged between stakeholders.

The key transformational knowledge points include the following:

- The trusting and financial relationships agriculture professionals have with agriculture producers are vital to acquiring landscape and management data.

- The aptitude of agriculture professionals is aligned with assessing natural resources.
- The production data sets collected by these professionals are applicable to calculating multiple sustainability indices.
- The additional costs associated with generating index-based natural resource assessments were minimal as long as the agriculture professional was engaged in the production aspect of the farm.
- Index-based accounting by geo-referenced points created scalable landscape intelligence.
- Index-based accounting enabled efficient data transfers to multiple sustainability stakeholders.
- A *shared governance* model is an efficient means to acquire data and share the values.
- Index-based portfolios act as a common platform for stakeholders to interdependently value natural capital and its outputs.
- The concept of an index-based assessment and valuation platform resonated as a potential solution to various agriculture systems in multiple countries.
- With these insights, a rich picture consisting of this target knowledge is created to show how the agriculture landscape system *ought to be*.

6.6   The system as it ought to be

To seek a resolution of a wicked problem, a rich picture of what the system *ought to be* is drawn as a companion to the system as it *is* (Williams and Van 't Hof, 2014). The rich picture in Figure 6.1 is based on the same components and stakeholders shown in Figure 2.3. The system as it *ought to be* is designed to capture the insights of the transdisciplinary story. Agriculture agronomists and other production professionals are engaged with both commodity and ecosystem service production and data management. The index values are connected spatially and temporally to the landscape to create scalable landscape intelligence and can be incorporated into sustainability asset portfolios at multiple scales.

Beyond just simply listing assets, these portfolios create new values. This causes new opportunities to emerge as participants are able to link assets across organizational and political boundaries (Morrison, 2013). With access to sustainability portfolios, utilities, government agencies, corporations, and nongovernmental organizations (NGOs) may procure the sustainability data they need to substantiate their sustainability claims. This process begins to align ecosystem service values in a similar manner to the way traditional commodity production values are aligned, where market, government, and NGO activities are complementary and

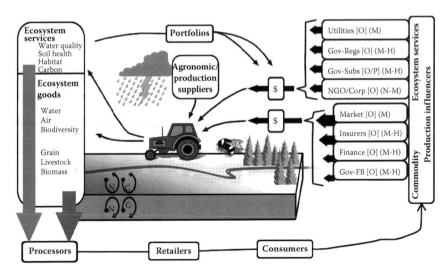

Figure 6.1 A rich picture of the system as it ought to be is based on the same boundaries and components of the system as it *is* in Figure 2.3. For this transformation to occur, the ecosystem services production influencers adopt an outcome-based accounting system to create an alignment of values similar to the commodity pathway. A common governance style is not mandatory, but a shift toward market governance may occur by adopting outcome-based accounting systems. A ecosystem service portfolio allows these less tangible assets to be exchanged.

additive. This begins to address the wicked problem of low and diffuse values and high and multiple transaction costs associated with exchanges for ecosystem services.

6.7   *Identifying the wicked problem and solution partners*

Seeking solutions to wicked problems often brings one into unchartered territory. After two decades of business research projects, Camillus (2008) stated that it is impossible to find solutions to wicked problems. Rittel and Webber (1973) state beginning down the wrong path may not only lead to failure, but cause additional problems. Camillus (2008) suggests that the simplest techniques are often the best, and [practitioners] with their tacit knowledge and commitment, are crucial in developing strategies to cope with or resolve wicked problems. Communication networks must cater to the flow, processing, and management of data, since information is necessary for all decision makers to participate.

Rittel and Webber (1973) suggests that one must use intuition to begin the process of resolving wicked problems and indicated that finding the problem is the same as finding the solution, of which both depend on the information gathered based on one's idea for solving it.

Using Rittel and Webber's suggestion of intuition and duality thinking, a conclusion is that the wicked problem of agriculture sustainability is the low and diffuse values and the high and multiple transaction costs associated with an ecosystem services *market*. Within this logic, the wicked solution is to *combine* the low and diffuse values associated with ecosystem services and *reduce and share* the high and multiple transaction costs. The solution is envisioned in Figure 6.2 as the inverse of the problem.

The three sources of the wicked agricultural landscape sustainability problem are identified as (1) varied scope and scale of natural capital outputs, (2) disparate stakeholder values, and (3) conflicting governance styles. These sources manifest into low/diffuse values and high/multiple transaction costs. Within this context, the three principle resolutions proposed are to (1) integrate natural capital data, (2) align sustainability actions, and (3) enable a shared governance platform.

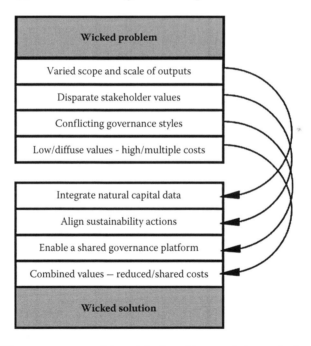

Figure 6.2 The components of the wicked problem and the wicked solution are shown to have a reciprocal relationship. The three principles of the wicked problem of agriculture sustainability lead to low and diffuse values for ecosystem services and high and multiple transaction costs. This is resolved with three wicked solution principles that result in combined values and reduced and shared transaction costs.

6.8   Wicked solution sources

Resolving the wicked problem of agriculture sustainability requires solutions from several governance sources. The source of technical, scientific, political, and wicked problems was described in Figure 2.1 with each issue emerging due to a combination of consensus and/or disagreement on values and knowledge. In review, when there is consensus on values and knowledge, the problems are technical; when there is consensus on values and disagreement on knowledge, the problems are scientific; when there is consensus on knowledge and disagreement on values, the problems are political; and when there is disagreement on values and knowledge, it is a wicked problem (Meuleman, 2013).

Nickerson and Zenger (2004) concluded that particular governance styles are more apt to solve specific types of problems. In their research of business firms and governance styles, they identified three governance styles: authoritative hierarchy, market, and consensus hierarchy. These three governance styles of business firms parallel the hierarchy, market, and network governance styles, respectively. To create a complementary solution matrix, Nickerson and Zenger's (2004) research on governance styles and problem solving was overlaid with Meuleman's (2013) typology of problems. Combining these problem and solution source concepts creates the associations shown in Figure 6.3. Hierarchical governance, relying on order, aligns with solving technical problems. Market-based governance, relying on innovation, aligns with

Values / Knowledge	Consensus	Disagreement
Consensus	**Hierarchy** (technical)	**Network** (political)
Disagreement	**Market** (scientific)	**Shared** (wicked)

Figure 6.3 The typology and source of problems and solutions are based on combinations of consensus and disagreement of values and knowledge. Based on the types and sources of problems shown in Figure 2.1, governance styles that are best adapted to address certain types of problems are paired with those problems. Hierarchy governance is suited for technical problems, market government for scientific problems, network governance for political problems, and shared governance for wicked problems. (Adapted from Meuleman, L., *Transgovernance: Advancing Sustainability Governance*, Springer, Heidelberg, Germany, 2013).

solving scientific problems. Network governance, relying on horizontal communication, input, and trust, aligns with solving political problems.

To complete the solution matrix, a shared governance concept is associated with resolving wicked problems. Since a shared governance model does not disregard other governance styles, the information gathering and problem-solving benefits of each governance style may be realized.

Shared governance benefits can be captured with an interactive platform.

Patterson et al. (2015) see supportive platforms and networks as particularly beneficial for developing thinking and capacity for transformations toward sustainability. Operating at different scales and with different sets of participants, platforms, and networks can offer different strengths, yet contribute in several common ways. This may lead to building *communities of practice* where participants are involved in sustained engagement and dialogue around particular issues of mutual interest.

6.8.1   Shared governance platform

The role of a shared governance platform is a similar role that firms play in traditional markets.

Williamson et al. (1991) explain a firm exists because it reduces the transaction costs that emerge during production and exchange, and can create efficiencies that external entities or individuals cannot. Due to the complexity of agriculture landscape sustainability, no one firm or entity can capture all the knowledge needed, and a solution demands more coordination than a typical firm. A shared governance platform is proposed as the means to capture the multiple and often less tangible values from the agricultural landscape and enable people to interact.

6.8.2   Enabling communities of practice

The purpose of a community of practice is to provide a way for practitioners to share and acquire information and provide support to each other to achieve their objectives. The intention of the shared governance platform is to create these connections, but also to provide the technological architecture to calculate and account for ecosystem services and enable transactions.

To do so, the shared governance platform uses land management indices that are able to *package* an extensive amount of data into singular units. It houses a cache of indices that together will function as a *landscape language*. The new language provides the form and content to create a useful method to communicate sustainability values across sectors and disciplines. A new language is vital to transcend the issues of the system as it *is*.

A more fluent communication of outputs, outcomes, and values would enable stakeholders from any sector to coordinate activities in a mutually beneficial manner. This allows *governance actors* to interact to achieve outcomes rather than depending on organizations with conflicting governance styles to communicate through multiple organizational layers. *Governance actors*, discussed in Chapter 8, are the individual representatives from the public policy maker, private policy maker, public practitioner, and private practitioner sectors. Aligning the sustainability activities among governance actors enables the shared governance objective of creating partnerships, equity, accountability, and ownership at the point of services (Porter-O'Grady, 2001).

The next three chapters examine how an index-based landscape language (Chapter 7), an alignment of sustainability activities (Chapter 8), and a shared governance platform (Chapter 9) contribute toward enabling a *wicked solution* to transform the agriculture system as it *is* to the system as it *ought to be*.

chapter seven

A landscape language

Communication is a key issue for solving wicked problems (Camillus, 2008). Language and communication have often been cited as barriers to interdisciplinary research (Stock and Burton, 2011). Such a scenario was illustrated in Chapter 3 with the various definitions of ecosystem services. These communication issues are exasperated as the number of individual sustainability stakeholders increases. Relationships become strained by a mutual lack of understanding of each other's goals and expectations relative to agriculture sustainability (Stokols, 2006).

Landscape indices can be used to bridge this communication gap and address the difficulties of directly measuring ecosystem outputs and outcomes at the landscape scale. Direct measurement is clearly impossible in the context of many of the most pressing questions of societal relevance, such as those regarding the impacts of climate change, large-scale agricultural intensification, and habitat loss. In these cases, indices are used as *measurements* to provide information on ecosystem structure, functions, and ecosystem service delivery (Stephens et al., 2015). Ecological indices have become an important topic of landscape research, as hundreds of indices are researched for landscape management and planning purposes (Uuemaa et al., 2009).

In this text, indices are proposed as the foundation for an agricultural landscape sustainability *language*. A common language is an important step to creating a more inclusive dialogue to overcome communication barriers. To be an effective communicator, an index should reduce a large quantity of data down to its simplest form while retaining essential information for the questions that are being asked of the data (Ott, 1978). In short, an index is designed to simplify. Landscape indices *package* complex data sets into a unitless number to provide information and insights related to land management in a succinct form (Bell and Morse, 2008).

In this form, indices can act as a common language that describes resource-management outcomes of the landscape in a succinct and transparent manner. Communication among government, environmental groups, conservationists, agricultural organizations, and farmers can occur relatively quickly at a scope and scale not possible with practice-based accounting. A functioning landscape language would include, as all functioning languages do, form, content, and use (Bloom and Lahey, 1978).

- *Form* refers to the structure of language. This includes the sounds, words, structures, and rules of sentences. The structure of a landscape language would be made of data points, data collection processes, index development, and calculations.
- *Content* refers to the meaning of the language overall, including the individual words that are used and the meaning created when these words are combined. The content of the landscape language is created by what the indices describe, such as the condition of the natural capital, the processes, and functions and its output. This content, referred to as landscape intelligence, may be spatial and temporal specific and scalable. Content consists of individual indices and portfolios.
- *Use* refers to the purpose of the communication, including why communication occurs. The purpose of an index-based landscape language is to transcend disciplinary and sector boundaries by supporting commonalities in defining, measuring, accounting, valuing, and exchanging values.

A primary purpose of creating a functioning landscape language is to create a common definition for ecosystem services. The four definitions from Chapter 3 are included: (1) the ability to relate MEA's ecosystem services to human use issues (Alcamo and Bennett, 2003), (2) align the definition with existing market definitions of goods and services (Brown et al., 2007; Daily, 1997), (3) recognize these values as spatially based (Luck et al., 2009), and (4) ensure relevancy within the broader economic system (Boyd and Banzhaf, 2007). Each of these attributes contribute to the overall goal of creating compatible aspects of accounting, measurement, ownership, valuation, and enabling transactions of ecosystem service assets.

7.1   Index-based language form

The structure of a landscape sustainability language should identify landscape data needs, data collection methods, types of indices, and index calculations or algorithms. It is analogous to a typical language consisting of sounds and letters, rules on how words are created, specific word types, and definitions. The form should create the basis to be able to "speak" to a broad audience on what is occurring on the landscape.

7.1.1   Landscape data

The landscape data needed to calculate index values includes the physical data of the landscape such as soil, topography, and vegetation, the management data, and environmental impacts associated with weather and various species. The number of data points can be numerous, but there is a limited amount of data to be collected from the landscape. For example, the AgEQA project required approximately 50 data points to calculate the five indices in the assessment portfolio, with each index requiring a different data subset.

7.1.2 Data collection

Data was collected using three primary methods: remote sensing, geographical information systems (GIS), and a participatory approach. This data was applied to a smart assessment process allowing data to be manipulated and to *flow* from the landscape upstream to those seeking sustainability values. To support an index-based language, each of these data sources must assess the landscape using smart methods, that is, the goal is to collect and compile the data in a format that creates new knowledge, and transfer this data and knowledge up the entire data supply chain to support each of the stakeholders as they seek and add sustainability values.

- Remote sensing is the use of satellite imagery, aerial photographs, or light detection and ranging (LiDar) to assess the condition or existence of land use or land-use responses such as the extent of forest cover and cropland, or the level of productivity of those land uses. LiDar is a remote-sensing system used to collect detailed topographic data. Remote-sensing techniques are useful for covering large tracts of land and can determine physical states as well as biophysical and chemical functions such as evapotranspiration and photosynthetic rates.
- A GIS is any system that captures, stores, analyzes, manages, and presents data linked to a location. Technically, a GIS is a system that includes mapping software and its application to remote sensing, land surveying, aerial photography, mathematics, geography, and tools that can be implemented with software that spatially maps and analyzes digitized data. This software can analyze temporal changes in ecosystems that correlate trends in ecosystem services with land-use changes. These data maps can also overlay social and economic information with the ecosystem information.
- A participatory approach is a method relying on farmers, stakeholder groups, resource managers, agroecological professionals, and scientific experts to supply resource information. It can be highly valuable as this approach collects knowledge not currently available in scientific literature, government agencies, and institutions. This information adds new perspectives, knowledge, and value to the assessments. The participatory approach is also the most challenging as there may be significant differences in the capacity of the individuals supplying the information and the costs to obtain it.

These data collection strategies can also be combined with each other to provide a more extensive measurement and analytical system. For example, the participatory approach could include gathering information from farmers on how they manage their production and natural resources. If this information is incorporated into indices within a common platform, then GIS and remote-sensing data could complement these participatory assessments.

7.1.3   Smart assessments

To create the basis for an effective landscape language, the data must become part of a *smart* assessment process that readily incorporates participatory data into GIS databases and remote-sensing data. According to the MEA project, a good assessment must be scientifically credible, politically legitimate, and useful for responding to the needs of decision makers. The assessment framework should be flexible in that the data is scalable, applicable in multiple manners, and creates knowledge to increase the resource-management capacity of the stakeholders (Alcamo and Bennett, 2003). A smart assessment begins at the landscape and is able to carry this data throughout the accounting, valuation, and transaction processes. It incorporates academic and nonacademic participants in the process of identifying sustainability values and creates a path toward interdisciplinary and transdisciplinary approaches.

Unsurprisingly, assessments focused on smaller areas or utilizing data collected at greater spatial resolution are capable of detecting greater variability than coarser, large-scale studies (Stephens et al., 2015). This local-level knowledge was explicitly valued by the MEA as it provides information that is often not documented by the science community. In the agriculture landscape, this local-level knowledge would consist of daily and seasonal activities that may impact the resources.

In the AgEQA case, the agronomist and farmer became the on-the-ground participants that brought new landscape knowledge to the watershed community. The indices were chosen from government agencies and scientific institutions and combined with GIS and remote-sensory production data. This smart assessment data became part of the farmer's ecosystem service portfolio that could be applied toward local, regional, and national objectives defined by government, NGO, and industry entities. The smart assessment captures the data and provides the path to transfer the data throughout value streams. These values are based on the index calculations.

7.1.4   Index calculations

Index calculations are based on specific data related to landscape features, management activities, weather, and environmental effects. For example, in the Agriculture Water Quality Certainty Program (AWQCP) case study, a water quality index (WQI) developed by the NRCS is calculated using data points such as field slope, soil hydrological group, soil erosion factors, soil organic matter content, nutrient management, tillage practices, weed and pest control, water management, and conservation practices.

Examples of other landscape sustainability indices include soil conditioning, habitat, and air quality indices. These indices represent the processes and functions of ecosystems that produce clean water and healthy soils, and sequester carbon. The number and types of environmental indices that can be developed is nearly limitless, with each potentially having multiple purposes and uses (Ott, 1978; Cvetkovic and Chow-Fraser, 2011; Uuemaa et al., 2009). Therefore, to support a landscape language, a smart assessment needs to have the capabilities to collect and compile data in a manner that enables index calculations with relative ease. The language form and structure creates the basis to be able to "speak" to a broad audience and relay the content of the language.

7.2 Creating language content

Content refers to the information to be conveyed by the language. In the case of landscape sustainability, the content of the landscape language is created by the data and the landscape indices that describe the condition of the natural capital: the processes and functions of ecosystems and their output. The language content creates *landscape intelligence*.

7.2.1 Landscape intelligence

Landscape intelligence is a similar, but broader term than *conservation intelligence,* which is described as described by Cox (2005) as the most up-to-date information about how landscapes are being managed for the purpose of directing government policy and programs.

Landscape intelligence is informed and guided by landscape properties that have only recently become available. Fulton (2011) defines landscape intelligence as an organizational and infrastructural layer to be used to develop new, alternative social–ecological conditions. It is required for the cultivation of more creative, ecological, and usable landscape practices better suited to the wicked problems faced in the twenty-first century.

Landscape intelligence begins with the collection and compilation of landscape data within a smart assessment process, applying this data to simple and compound indices, and compiling this knowledge into portfolios.

7.2.2 Simple and compound indices

If indices are viewed as the words of landscape sustainability, then simple indices describe a one-dimensional aspect of the landscape, and compound indices describe the combined effect of multiple dimensions of sustainability.

Simple indices calculate for just one component, such as a WQI. As a simple index, the WQI only shows the reader if the water quality meets a certain standard, and does not directly show the reader what actions are needed to improve the water quality or what is causing the particular outcome. Someone familiar with the index equation and the scientific principles included in the calculations would be able to identify actions that may increase or decrease the index score. Simple indices cannot illustrate the effects or the comparative value that the one component has relative to other measurements.

Compound indices combine simple, one-dimensional indices into a measurement capable of addressing many (environmental) issues at once. It allows aggregation and integration of dissimilar indices to present data into a tool capable of accommodating multiple data sets, adjusting weights, and being used to represent multiple stakeholder interests. The compound index is generally constructed so that the data sets are fixed and the weighting of values is implicit. In other words, the developer of the index inserts weighting and values within the structured index, rather than allowing the user to define these characteristics.

An example of a compound index is the Environmental Benefits Index (EBI) of the United States Department of Agriculture (USDA) Conservation Reserve Program (CRP). It contains five environmental components as well as a cost variable. The EBI functions as a static instrument to determine program eligibility for a particular year, but over the years, the parameters and weighting can be and are adjusted to accommodate new scientific findings, social interests, and political desires. Simple and compound indices represent ecological values that can be compiled into a portfolio representing natural capital asset management.

Indices represent compromise designs, involving trade-offs between ease and cost-of-data measurability, scientific validity, transparency, and relevance to users. At best, they are "optimally inaccurate" (OECD, 1999). They are representations of reality and do not have to be technically perfect (Bennun et al., 2014). This does not mean that indices have no value for measurement: rather their value depends on how they are applied and interpreted.

7.2.3   Natural capital asset portfolio

A natural capital asset portfolio is a grouping of agricultural landscape sustainability values. It is an extended use of the smart assessment. Table 7.1 is an example of the data associated with an agriculture landscape portfolio. Individual management units, such as fields, are the basis for

Table 7.1 A natural capital asset portfolio compiles index values representing
ecosystem service outcomes of a landscape

Field nos.	101A	102A	103B	104B	Totals	
Hectares	80	120	100	300	600	
Index					**Range**	**HWA**
SCI	88	85	68	92	68–92	86.0
WQI	89	88	70	78	70–89	80.1
HSI	72	62	63	68	62–68	66.5
SLI	82	77	64	80	64–82	77.0

the index calculations, and are aligned with the Luck et al. (2009) service
providing unit for ecosystem services. Three simple indices (soil conditioning, water quality, and habitat suitability indices) and one compound
index (sustainable landscape index) are used as examples of potential
portfolio values. The three simple indices are actual USDA-developed
indices and the SLI is a fictitious index used to illustrate the use of a compound index. (An EBI would be a comparable compound index lands not
actively farmed, such as restored to a prairie setting.)

Simple indices for soil, water, and habitat (soil conditioning index [SCI],
WQI, and habitat suitability index [HSI]) account for individual resources,
and a compound index (SLI) accounts for the sustainability status. In
this portfolio, the value of each field, the range of scores, and the hectare-
weighted averaging for each field and for the entire operation are provided

The portfolio provides additional content to the landscape language
by connecting landscape units to index values and expanding the landscape intelligence. The hectare-weighted average (HWA) provides an
index value relative to the overall landscape assessed.

The indices express the *ecological values* of the portfolio. The *economic
value* is relative to the price sustainability demanders are motivated to
pay. The portfolio expresses the intrinsic value associated with landscape
management. The purpose and use of the portfolio is to identify and communicate the extrinsic value as it relates to societal sectors and the economic system at large.

7.3 Purposeful uses

The purpose of an index-based landscape language is to allow stakeholders to communicate the varied outputs and outcomes of natural
capital in a far more productive manner than currently occurs in the
system as it *is*. The usefulness is for stakeholders to identify common-
alities and begin negotiations to achieve their sustainability objectives.
A common language can transcend disciplinary and sector boundaries by allowing stakeholders to seek out and create synergies within

the broad boundaries of landscape sustainability. By finding common-alities in how they define, measure, account for, value, and conduct transactions, stakeholders can begin to combine the low and diffuse values of ecosystem services and reduce and share the high transaction costs.

7.3.1   Another market marvel?

Economist Hayek called the free market system a marvel because just one indicator, *price*, spontaneously carries so much information that it guides buyers and sellers to make decisions to obtain what they want (Gwartney, 2010). Of course, this *marvel* is much easier to create when the commodity can be directly measured in kilograms and liters. But like the numerical indicator of price that reflects thousands of decisions made by people who do not know each other, a landscape index reflects the decisions, activities, and conditions of a particular parcel as it relates to sustainability criteria. A landscape index has functions similar to price.

First, the function of price is to enable commerce to be coordinated by transmitting information about available resources and the demands placed on them. Second, it provides incentives for people to adopt the least costly methods of production and to use available resources for their most highly valued uses. And third, prices determine how much income is distributed and to whom (Friedman, 1981).

By combining the attributes of these two indicators (price and index), a variety of uses related to assessment and valuation are possible. Beginning with a smart assessment, values can be attributed throughout the assessment, planning, assurance, and transactional processes.

7.3.1.1   Useful applications

A single index is able to address numerous aspects of the valuation chain beginning with the smart assessment and continuing through the valuation and transactional aspects of landscape sustainability. Nine useful aspects include the following:

1. Assessing of the condition of resources
2. Planning basis for legislators, agencies, corporations, and producers
3. Analyzing resource-management effects
4. Monitoring outputs and trends
5. Communicating resource conditions and trends
6. Educating the public, government, and producers about status and goals
7. Researching resource-management strategies and outcomes
8. Standard development for regulations, incentives, supply chains, and objectives

9. Valuation unit component for the following:
   a. Regulatory assurance
   b. Market access/participatory benefits
   c. Payment for ecosystem services
   d. Landscape intelligence data
   e. Liability/risk management
   f. Sustainability supply chain claims

Within the context of this language structure and content, ecosystem services can be defined, measured, owned, valued, and transacted.

7.3.1.2   Defining eco-services

To enable an index to be a *market marvel*, ecosystem services need a functional definition within the context of the ecological and economic systems. As noted in Section 3.3, the system as it *is* lacks a common definition for ecosystem services. This is a primary obstacle for stakeholders to communicate their sustainability needs and express values (Boyd and Banzhaf, 2007).

For this text, natural capital outputs and outcomes will be defined separately as ecosystem goods *and* ecosystem services and will be referred to as ***eco-services*** to differentiate this definition from other terms. *Eco-services* will be based on the definition used by Daily (1997) and Brown et al. (2007). In this context, *eco-services* are defined in the traditional economic sense of a service being an intangible, time-dependent process often associated with a location or site conditions. *Eco-goods* are defined in the traditional economic sense of a good as a tangible, consumable item.

This definition is inclusive relative to the other important attributes associated with human needs, spatial connections, and economic indicators. Since these eco-services are accounted for within a geo-referenced, index-based platform, it can adopt the service providing unit (Luck et al., 2003) concept to address location-specific values. This same accounting structure could create the connection to the broader economic metrics promoted by Boyd and Banzhaf (2007) and include FEGS and BRI values. From the perspective of the MEA definition, eco-services would capture the definition of *regulating* ecosystem services.

These eco-services become a *geo-referenced index package* and are referred to as a natural capital unit (NCU). The NCU acts as a commodified natural capital asset to simplify measurement, ownership, valuation, and transactional processes. The NCU commodifies the management of ecosystems as place-based outcome.

7.3.1.3   Natural capital units

NCUs are the tradable units of eco-services. They are calculated using indices applied to a natural capital cell (NCC), a specific area of land. The

NCU represents the functions and processes within the NCC. The NCU, an index calculation and location, becomes the unit of eco-service ownership. This is a key feature to enable trade and commerce to support agricultural landscape sustainability. The NCU *converts* what is typically and traditionally viewed as a nonexcludable and nonrival public good into an excludable and rival private good, or an excludable and nonrival club good. This conversion is further explained in Chapter 9.

Once ownership is established for NCUs, the value of eco-services will be based on the interaction and negotiation between an NCU buyer and seller. A person or entity may be interested in purchasing NCUs to substantiate a sustainability claim, to promote a level of landscape sustainability, or to achieve some of their sustainability objectives for a watershed or a particular product. Regardless of the motivation, NCUs are generated within the context of landscape data and how the landscape is managed.

7.3.2　Mapping earth's factory floor

NCUs express the viability of earth's *factory floor*: its landscape and its capacity to generate outcomes and outputs of natural capital. Costanza et al. (2014) estimated the value of the earth's natural capital outcomes and outputs for oceans and terrestrial environments at $124 trillion/year (Table 7.2). This value exceeds the global gross domestic product (GDP) of $75 trillion. These estimates include values from two marine biomes:

Table 7.2 Natural capital values (in US$) produced each year relative to biomes (in trillions of US$) and hectares

Biome	Area	Values	
	ha e6	$T	$/ha/yr
Marine	36,302	49.7	1,369
Open ocean	33,200	21.9	660
Coastal	3,102	27.8	8,962
Terrestrial	15,323	75.1	4,901
Forest	4,261	16.2	3,802
Grass/rangelands	4,418	18.4	4,165
Wetlands	188	26.4	140,426
Lakes/rivers	200	2.5	12,500
Desert	2,159	–	–
Tundra	433	–	–
Ice/rock	1,640	–	–
Cropland	1,672	9.3	5,562
Urban	352	2.3	6,534
Total	51,625	124.8	

open ocean and coastal waters with hectare values of $660/yr and $8,962/yr, respectively. The nine terrestrial biomes averaged $4,901/ha/yr with wetland eco-services valued the highest at $140,426/ha/yr and the low range of forests at $3,802/ha/yr. Insufficient data were available to estimate desert, tundra, and ice/rock biomes.

Some of this total is already included in GDP as marketed goods and services. But much of it is not captured in GDP because not all eco-services are marketed or fully captured in marketed products and services. Costanza et al. (2014) estimates that these eco-services (i.e., storm protection, climate regulation, etc.) are much larger in relative magnitude than the sum of marketed goods and services. This is possible as not all human benefits, economically and otherwise, are included in the GDP.

7.3.2.1 Natural capital values and GDP

For any economic value to be recognized at the national and global scale, it must be measured within the accounting system of the nation. The System of National Accounts (SNA) is a conceptual framework that sets the international statistical standard for the measurement of the market economy. It is published jointly by the United Nations, the Commission of the European Communities, the International Monetary Fund, the Organization for Economic Co-operation and Development, and the World Bank. The SNA consists of an integrated set of macroeconomic accounts, balance sheets, and tables based on internationally agreed concepts, definitions, classifications, and accounting rules. Together, these principles provide a comprehensive accounting framework within which economic data can be compiled and presented in a format that is designed for purposes of economic analysis, decision making, and policy making (UNSD, 2015).

Conventional measures of economic activity include GDP and net domestic product (NDP). Conventionally, GDP is constructed as a measure of the output of the market sector. In this manner, GDP has serious deficiencies in regard to natural resource stocks, whose use contributes to current income flows (Harrison, 1989). Although the "environment" is not completely invisible in national accounting, its treatment produces some curious results. Wright (1990) states that it often appears better, economically speaking, to cause environmental damage and then repair it than to avoid causing the damage in the first place; this is hardly an efficient form of economic growth.

The NDP equals GDP minus depreciation on a country's capital goods. NDP accounts for capital that has been consumed over the year in the form of housing, vehicle, or machinery deterioration. The depreciation accounted for represents the amount of capital that would be needed to replace those depreciated assets. If the country is not able to replace the capital stock lost through depreciation, then GDP will fall, unless the capital lost is natural capital.

A commonly cited example of the inappropriateness of the GDP to measure the value of the economic system is the Exxon Valdez oil spill of March 24, 1989, where 40.9 million liters of crude oil were spilled into the sea, covering 28,000 km² of ocean. Thousands of animals died immediately, including an estimated 500,000 seabirds, 1,000 sea otters, 300 harbor seals, 250 bald eagles, and 22 orcas, as well as the destruction of billions of salmon and herring eggs. Oil can still be found on the beaches of Prince William Sound today. Economically speaking, the accident generated an estimated US$5 billion in additional economic activity, much more than the straight delivery of the cargo would have produced (Jefferies, 2016).

In a more general statement, the shortfalls of GDP were clearly articulated by Repetto (1988) who stated that a country could exhaust its mineral resources, cut down its forests, erode its soils, pollute its aquifers, and hunt its wildlife to extinction, but measured income would rise steadily as these assets disappeared.

Harrison (1989) notes that GDP is "gross" meaning that it does not consider the depreciation of capital stock, such as the factories and infrastructure that produced the goods or the depreciation of natural capital. She states that first it may seem counterintuitive that the degradation of permanent resources leads to an increase in GDP. This reflects the common lack of awareness that the "gross" in GDP means before allowance has been made for consumption of capital.

This perspective gives a sense that the GDP is not a perverse system as it relates to natural capital. But the challenge still remains to have an NDP accounting system that can generate an economic allowance when the consumption, degradation, or improvement in the capacity of natural capital occurs.

7.3.2.2 Agricultural NCC values

To include natural capital values into the GDP or NDP, the NCUs must reflect natural capital appreciation or degradation at the landscape relative to the overall economy. Stoneham (2009) states this valuation is one-sided. If a landowner excludes livestock from part of the farm to allow habitat to regenerate, GDP will fall because the reduction in livestock production will be accounted for, but not the production of eco-services and/or the increase in the capacity of natural capital. Harrison (1989) states natural capital degradation and appreciation can be captured as annual changes and measured with air and water quality indexes.

To include these values at both local and global scales, the NCC is designed as a 9 m × 9 m land unit. It is at this scale that the production of small diverse farm operations can be captured as well as precision farming techniques and output from farms with thousands of hectares.

A 9 m × 9 m NCC is also small enough to delineate various-sized landscape features such as streams and wetlands. Digitizing the entire earth's surface using these 9 m × 9 m units yields approximately 6.4 trillion NCCs. Table 7.3 identifies the approximately 1.37 trillion NCCs covering the terrestrial biomes excluding desert, tundra, and rock/ice biomes. The average NCC value for each biome and the total value of each biome are listed. Agriculture would include the cropland and the grass/rangeland biomes for a total of 751 billion NCCs with a value of US$27.7T.

These totals include both marketed goods (food and raw materials) and nonmarketed goods and eco-services. For example, the value of a typical corn crop in an Iowa cornfield yielding 13.5 tons/hectare (200 bushels/acre) at US$ 178/ton (US$5/bushel) creates an approximate value of US$30 of corn per NCC. This compares to the estimated average value (Costanza et al., 2014) of the cropland NCC value of US$45.06/year in Table 7.3. In this simple calculation estimation, the nonmarketed eco-service value would be US$15 (US$45.06–US$30) per NCC or 33% of the total NCC value. In this case, the corn and perhaps the stover would be the marketable commodities, and the NCUs may consist of water, soil, carbon, and habitat values depending on the management activities.

It should be noted that Costanza et al. (2014) emphasized that valuation of eco-services is not the same as commodification or privatization and conventional markets are often not the best institutional frameworks to manage them. However, Costanza et al. (2014) state eco-services must be (and are being) valued, and new, common asset institutions are needed to better take these values into account.

Table 7.3 Terrestrial biome types, number of natural capital cells (NCCs) per biome (in millions), value in US$/NCC, and total biome values (in US$ millions)

Biome	NCCs e6	$/NCC/yr	Total value e6
Forest	526,049	30.80	16,200,891
Grass/rangelands	545,432	33.74	18,401,012
Wetlands	23,210	1,137.51	26,401,452
Lakes/rivers	24,691	101.26	2,500,137
Cropland	206,420	45.06	9,300,511
Urban	43,457	52.93	2,300,126
Total	1,369,259		75,104,131

Source: Costanza R. et al., *Global Environmental Change*, 26 (2014), 152–158, 2014, doi:10.1016/j.gloenvcha.2014.04.002.

Note: Total NCCs in the six biomes listed are approximately 1.4 trillion. Total market value is US$75.1T.

7.3.3   Disruption toward harmonization

The creation of a new NCU landscape language to communicate sustainability goals and to be used as an accounting and valuation platform for eco-services at local and global scales is a *disruptive innovation* strategy. Disruptive innovation is a term from Christensen's (1997): *The Innovator's Dilemma* describing a market force that occurs due to the introduction of a new process, technology, and/or product that appeals to a new customer base enabling them to approach an issue with a different set of values and strategies. Disruptive innovation can and does occur in many fields and sectors.

Kara Hurst, former CEO of The Sustainability Consortium, recognizing their sustainability vision remained largely unrealized, stated at the 2014 Green Biz forum that for sustainability to reach the next plateau *disruptive innovation* must occur (Hurst, 2014). To introduce disruptive innovations and achieve this goal, The Sustainability Consortium has developed science-based tools and methodologies to identify ways businesses can improve products and help companies answer the question of how businesses can hone in on where the real impacts are, and where they should collectively focus to make these improvements.

In many respects, the disparate stakeholder efforts in Chapter 4 and the case studies used to describe a transdisciplinary solution in Chapter 6 are examples of disruptive innovation strategies: the introduction of a new process, technology, and/or product that appeals to a new customer base enabling the innovators to approach an issue with a different set of values. Within the context of the NCU strategy, three global disruptive innovative strategies are reviewed: a global environmental mechanism (GEM) (Esty and Ivanova, 2002), a global environmental asset portfolio (GEAP) (Costanza et al., 2000), and developing a harmonized natural capital protocol (Maxwell et al., 2014).

7.3.3.1   Global environmental mechanism

Esty and Ivanova (2002) envision a GEM that draws on modern information technologies and networks to promote cooperation in a faster and more effective manner than traditional institutions. This network-based GEM would build on the functioning elements of existing institutions and create new structures where gaps exist. This global institutional mechanism would grow organically as consensus develops on issues and needs, but not as a new international bureaucracy.

Esty and Ivanova (2002) postulate a GEM would contain a data collection mechanism that calculates sustainability indicators and benchmarks. It could be used as a repository for information to provide continuous and transparent reporting effort. This data set could be used to assess environmental processes and trends, and assist forecasting long-term trends and environmental risks. It would have a means to facilitate transactions related to environmental issues in return for payment or achieving policy objectives. This mechanism could be used to establish policy guidelines for global commons and shared natural resources and provide businesses and NGOs the means to directly participate in problem identification and policy analysis. A finance mechanism could mobilize public and private resources to promote the best options suited to national and local solutions in coordination across all sectors.

Through these capacities, the GEM would contribute to addressing three institutional gaps: the jurisdictional gap, the information gap, and the implementation gap. Simply put, data can make the invisible visible, the intangible tangible, and the complex manageable. It is a re-engineering aiming for a new, forward-looking, and more efficient architecture that will better promote the environment while also serving governmental, public, and business needs. Its logic is based on a globalizing world requiring better and more modern ways to manage ecological interdependence (Esty and Ivanova, 2002).

7.3.3.2   Global environmental asset portfolio

The concept of a GEAP emerged in 2000 from a roundtable gathering of global experts. The discussion moved beyond the "environment as a debate" by focusing on the common view that the environment is a productive asset shared by all of humanity (Costanza et al., 2000). The roundtable suggested humans need to manage these assets at least as wisely as individual investors manage their stock portfolios, but recognized the fundamental problem with environmental management is that no effective institutions exist at the appropriate scale. An institution is needed for managing humanity's collective behavior and its common global environmental portfolio, rather than managing the earth assets as small, independent subunits, none of which had to account to any other for the resources it used.

The roundtable experts promoted the use of strategies by financial managers that work on a broad range of complex assets. The first rule of asset management is to protect the stock of assets that create wealth. In the context of the environment, the natural capital must be protected so that the flow of eco-services from the stock continues. The second rule is to hedge your investments, often referred to as "don't put all of your eggs in one basket." Relying on technology to solve all environmental problems is risky. Ecosystem preservation is a hedge against the possibility that technology alone will be unable to provide humans and their economy with adequate resources and eco-services. The third rule is insuring the asset. Insuring an environmental asset means one should not degrade its capacity to produce the goods and services humans rely on and should protect oneself against the worst-case scenario.

To manage the environment as an asset, Costanza et al. (2000) recommend an institutional framework for sustainable governance. This framework would promote responsibility for those with access to environmental resources; it would be able to match the scale(s) of the environmental issue to the relevant decision makers and enable adaptive management strategies. This framework would identify environmental costs and benefits and ensure they are appropriately allocated, and all relevant stakeholders would be engaged in the formulation and implementation of decisions concerning environmental resources.

7.3.3.3   *Harmonizing natural capital valuation*

The Natural Capital Coalition (NCC) is a global network of corporations, governments, and NGOs calling for a harmonization of natural capital valuation. It views the growing number of fragmented natural capital activities as a jigsaw puzzle with some of the pieces in place, but mostly disconnected and gaps needing to be filled. This is causing confusion for businesses and investors trying to make informed decisions on risk mitigation, securing resource supply, creating long-term value, resilience, and profitability (Maxwell et al., 2014).

The NCC seeks ways to fill technical gaps relating to natural capital indicators, data, and classification systems to facilitate the mainstreaming of natural capital valuation and emerging environmental economic accounting metrics. The framework would incorporate harmonized principles for what should be measured and valued, and how. This would include clarity on types of capital and the connections between natural, financial, manufactured, societal, human, and intellectual capitals. The NCC state market push and pull factors are needed to motivate behavior change to integrate natural capital and transform business models (NCC, 2015). At the November 2015 World Forum on Natural Capital in Edinburgh, the NCC launched the draft Natural Capital Protocol and accompanying sector guides. The purpose of the draft protocol and guides is to help businesses systematically integrate their relationship with nature into their strategy and operations. In 2016, the NCC participants will provide feedback on how well the categorization scheme works for disparate entities relative to indirect impacts on natural capital and the direct impacts of supply chains.

7.4   *Imagining a common landscape language*

Each of the three global approaches requires a new, fundamental language of sorts to communicate resource concerns, transmit and compile data, and exchange values among sectors and entities at local and global scales. It is this daunting scenario that perhaps convinced Hurst to state the system as it *is* requires disruptive innovation to move it to the next plateau. Brown (2010) states that the ingredient needed to transcend the existing system of landscape sustainability is *imagination.*

Esty and Ivanova (2002) imagined a networked GEM, while Costanza et al. (2000) envisioned an institutional framework to manage the complexity of an environmental portfolio; the NCC participants are imagining an accounting system for their natural capital impacts. Imagining an index-based language based on a digitized earth-scape consisting of 6.4 trillion NCCs is a step toward a potential transdisciplinary solution. This global, index-based language could address Conklin's (2006) issue of fragmentation, referring to the multiple perspectives, understandings, and intentions of sustainability stakeholders.

This fragmentation results in a lack of adherence to ensuring scientific principles (Naeem et al., 2015). Of 118 payment-for-ecosystem (PES) projects examined by 45 scientists and practitioners from government, nongovernment, academic, and finance institutions, only 40% met the four principles to ensure scientific integrity. Naeem et al. in the March 2015 *Science* issue stated baseline data, monitoring of key environmental factors, and services, recognizing that ecosystems are dynamic, and inclusion of metrics are key to ensure the success of sustainability projects.

Brown (2010) stated that as a wicked problem, it is highly unlikely that the many interests involved would be willing or able to work together. These interests need an open-ended and collective framework that stretches their imagination to include the contributions of each other. This collective framework consists of land management indices as the basis for the form and content (the words and intelligence) of a landscape language. These new terms within the context of a digitized landscape support the use of NCUs as potential *market marvels.* Similar to how *price* aligns market participants, indices can align governance actors and market participants.

Aligning the sustainability conversations is a necessary step to begin the alignment of the stakeholder activities that contribute to and achieve common sustainability objectives. The following chapter illustrates how organizations, in their sustainability efforts, are becoming more inclusive in their conversations, their sustainability activities, and their organizational governance styles. This trend can be accelerated with a common landscape language.

chapter eight

Aligning sustainability activities

Disparate stakeholder values, the second wicked problem component, is resolved by aligning the sustainability activities of the stakeholders. As discussed in Chapter 4, a variety of activities are conducted independently with mixed and unaccountable results. Since no one entity is able to dictate order and process in complex social systems, an alignment of stakeholders must occur somewhat organically and emerge from the interaction of participating stakeholders.

To achieve this alignment, a more flexible approach in decision making is needed to create an ongoing process of negotiation (Sayer et al., 2013). Rigid organizational governance styles must play a less prominent role in guiding stakeholder actions (Marsh, 2000) and a more systematic and seamless approach including dissolving or redrawing traditional disciplinary and sector boundaries is needed (Haines, 2009). These boundary transformations are occurring in businesses, political parties, and public organizations resulting in an increase in collaboration among private, public, and nonprofit organizations (Avril and Zumello, 2013).

Avril and Zumello (2013) call for researchers to describe and analyze these new models of governance to help us understand their impact on both citizens and traditional institutions. This call is addressed, in part, by assessing three groups of case studies. The first group is a comparison of the two eras of the United States Department of Agriculture's (USDA) conservation delivery system (CDS), the second group consists of three state-level (Minnesota) agriculture environmental assurance pilot projects, and the third group consists of six emerging efforts addressing agricultural landscape sustainability from several sectors at regional and national scales in the United States.

The purpose of reviewing these case studies is to (1) identify the *governance actors* and their roles, (2) describe *shifts* in organizational governance, (3) describe the governance frameworks, and (4) plot the organizations' *governance footprints* relative to their accounting strategies and predominant governance styles. Understanding these shifts and the role of key actors is seen as a critical component of enabling transformation and adaptive capacity within social–ecological systems (Patterson et al., 2015).

8.1   Governance actors

Governance actors are individuals with the capacity to make decisions and conduct activities to achieve specific or desired outcomes. Governance actors are categorized as either public or private and either policy maker or practitioner.

A *governance compass* (Figure 8.1) was created to illustrate four aspects of governance: public, private, policy maker, and practitioner. The four aspects of governance create four sectors representing *governance actors*: public policy makers, private policy makers, public practitioners, and private practitioners.

Viewing the governance compass from the agriculture sustainability perspective, governance actors include the following:

- Public policy makers: Elected legislators and local, state, and federal government policy staff
- Private policy makers: Corporate and nongovernmental organization (NGO) policy staff related to sustainability supply chains, environmental and conservation objectives, and social benefits
- Public practitioners: Federal, state, and local government and public educator agents providing technical assistance for conservation practices, landscape planning, and field assistance
- Private practitioners: Farmers, agronomist, and private sector professionals directly involved in providing and using equipment, products, and services for the management of agriculture landscapes

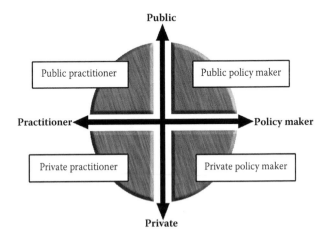

Figure 8.1 The four aspects of a governance compass (public, private, policy maker, and practitioner) create four governance actor sectors: public policy maker, private policy maker, public practitioner, and private practitioner.

Each category, particularly those nontraditional, nonstate actors, need to be recognized as significant contributors as they often have the knowledge needed to deal with complex social–ecological systems (Loe et al., 2009). These private practitioners are often the *target* of sustainability policies, and so they are the actors able to provide the outcomes and data that represent the policy objectives of the public and private policy makers. Understanding each sector's place in the governance compass is a critical step in aligning their activities. Governance actors need to understand their roles, responsibilities, and relationships among other actors to ensure expertise, skills, and abilities are utilized in an efficient manner. The governance compass provides an organization framework to describe the existence and location of each actor sector and to illustrate potential relationships and roles.

8.1.1 Shifting actor roles

Participation of nonstate actors in the process of governance initiates the transition from government to governance (Loe et al., 2009). It is through these new governance roles that actors shape new courses to coordinate actions and achieve their objectives (Lowndes and Skelcher, 1998). Private environmental governance is an increasingly important aspect of environmental law and policy, worthy of attention from policy makers and practitioners, in that it offers new responses to some of the most intractable environmental problems. Environmental policy makers, often assumed to reside only within government, may be a corporate chief financial officer, university procurement official, or the chair of the board of a private organization (Vandenbergh, 2014).

Understanding governance as a collaboration of actors, rather than organizations, supports a strategy to align activities of the actors rather than policy strategies to address the inherent conflicts associated with organizational governance styles. Dissecting governance to this actor level identifies the nuances of stakeholders' roles and relationships and reveals ongoing *shifts* in governance styles.

8.2 Governance shifts

Organizations seldom rely strictly on one style of governance. In reality, organizations are characterized by a combination of the three governance styles, although one style often dominates the organization structure (Dijk and Winters-van Beek, 2009).

As the world becomes increasingly decentralized and social and economic boundaries blur, the traditional top-down hierarchy governance is less appropriate (Roe, 2013) and market governance may not represent all stakeholders (Fábián, 2010; Dijk and Winters-van Beek, 2009).

This decentralization and diversification have caused the traditional governance models to destabilize and new governance arrangements to emerge (Kersbergen 2004). The predominant trend is to move away from hierarchy and market models toward network styles (Roe, 2013). This shift from *government to governance* is seen on all scales from the local, state, and regional governments to national states and on to various forms of inter-governmental arrangements at the international and global levels (Jessop, 2002). Dijk and Winters-van Beek (2009) describe this as a turning government, meaning it is leaning toward more horizontal modes of governance rather than vertical hierarchies.

External factors are one cause of these governance shifts. Threats to ecological and social values and scarcity of resources mobilize individuals to create networks, causing a shift in governance (Chaffin et al., 2014). As society becomes more complex and fluid, new methods of governing are necessary and new governance models are devised to rely less on hierarchical institutions. Governments are moving away from direct service provision toward an enabling role with an increased use of private organizations to deliver public services (Heywood, 2003).

This shift is also caused by the public's desire for more inclusive food systems and the need to not only know the "what and where" but the "how and why" of producing food (Sayer et al., 2013). In response to these new societal challenges, organizations realign and redesign their roles and governance strategies. These governance shifts are seldom a shift from one governance style to another, but an adoption of characteristics of other governance styles to create a governance mix, or a meta-governance strategy that addresses the evolving social sentiment.

8.2.1 Meta-governance

Meta-governance is the thoughtful mixing of market, hierarchy, and networks to achieve the best possible outcomes (Jessop, 2003). Meta-governance, in the definition of Meuleman (2008), is an approach which aims to design and manage a preference for a mix of institutions, consisting of elements of hierarchical, market, and network governance. Meta-governance is not a governance style, but an approach aiming at finding smart combinations of hierarchical, network, and market governance. It is a strategy to combine bottom-up, top-down, and horizontal networks into a particular framework to address a particular situation (Meuleman and Niestroy, 2015).

Meta-governance can operate in a context of *negotiated* decision making among several entities. Market competition will be balanced by network cooperation and the state is no longer the sovereign authority, as it becomes just one participant among others and contributes its own distinctive resources to the negotiation process (Jessop, 2002).

Meuleman (2008) sees meta-governance as a natural progression for governments, corporations, and NGOs to continue to address their missions while maintaining their structure and responsibilities. Figure 8.2 illustrates the meta-governance space among the three governance styles.

Meta-governance can also operate within an organization as it shifts away from hierarchical governance toward a more network governance and adopts new partnerships, alliances, agreements, and capacity building (Meuleman, 2013). These shifts in governance ratios or meta-governance strategies result from entities incorporating different governance traits: the building blocks of governance frameworks (Meuleman, 2008).

Meta-governance is a means to coordinate governance by combining specific traits of hierarchical, market, and network governance styles. Fleming and Rhodes (2004), among others, recognized the value of a multiple governance approach and stated the future [solutions] will not lay with either market, hierarchy, or network but with all three. The trick will not be to manage contracts or steer networks but to mix the three systems effectively when they conflict with and undermine one another.

Sørensen and Torfing (2012) point to four recent conditions: the fiscal crisis, the proliferation of wicked problems, growing policy-execution problems, and low growth rates that have put pressure on governments to seek new meta-governance strategies to enhance the innovative capacity of the public sector. The driver in each of these cases is collaboration. Compared with other forms of governance such as hierarchies and

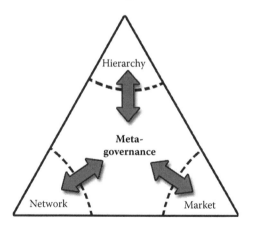

Figure 8.2 Meta-governance is illustrated as a strategy compiled from the characteristics of each of the three typical governance styles. The mixture contains various approaches to relationships, decision making, and compliance aspects. The meta-governance approach is in response to the governance trilemma described in Figure 5.3. (From Meuleman, L., *Public Management and the Meta-governance of Hierarchies, Networks and Markets: The Feasibility of Designing and Managing Governance Style Combinations*, Physica-Verlag, Heidelberg, Germany, 2008.)

markets, the form and composition of meta-governance can be adjusted so as to fit a specific purpose and occasion.

8.2.2   Meta-governors

The success of meta-governance as an arena for collaborative policy and service innovation is dependent on the approach of the *meta-governor* (Sørensen and Torfing, 2012). A meta-governor is someone who takes up some leadership role in governance analysis, design, and management (Meuleman, 2015). Meta-governance is not confined to one actor, and several meta-governors can exist at the same time (Kroemer, 2010). Meta-governors can reside in any sector (Rhodes, 1996), but they often originate from the public policy-maker sector as politicians and public administrators.

Successful meta-governors are those that are able to look beyond their own perspective and recognize hierarchy, market, and network elements as the building blocks to design and manage mixtures that will work well in certain contexts. They also need to have the ability to understand tensions and conflicts between elements of the governance styles (Meuleman, 2008) as their activities affect the network relationships, conflicts, and trust (Kroemer, 2010).

A meta-governor can use four approaches to guide network or meta-governance activities: design, framing, participation, and management (Sørensen and Torfing, 2012). Design and framing are hands-off approaches. Design activities address the structure, scope, composition, and procedures of a governance network. Framing activities are about formulating political goals and objectives to be pursued by the network. Participation and management are hands-on approaches with interaction with the governance network. They consist of decision making on overall objectives, policy processes, and goals. Management activities are aimed at increasing interaction and at strengthening relationships through mediation and by decreasing transaction costs (Kroemer, 2010).

Jessop (2002) stated that due to these options and challenges, meta-governors must be able to handle complexity very well and anticipate unexpected events. That is the rationale for adding a meta-governor to this complexity: to take a wider perspective on problem setting, possible solutions, and the choice of institutions, instruments, processes, and actor roles to provide leadership in developing multilevel governance frameworks (Meuleman, 2015).

8.3   Governance frameworks

A *governance framework* can be defined as the complete set of instruments, procedures, processes, and actors (and their roles) designed to tackle a particular issue (Meuleman, 2014). Building a governance framework begins differently for each entity as the preference for a governance style originates in their traditions, culture, political practice, and geography within the context of their environmental, social, and economic circumstances (Niestroy, 2005).

Meuleman (2008) compiled a list of up to 36 dimensions of governance and how each of the three styles addresses them. Table 8.1 includes 12 dimensions (of those 36) and the modes of addressing them. For example, in the *control* dimension, each governance style contains a specific method to engage with participants: hierarchy uses authority, market governance uses price, and network governance uses trust.

Preferences for one governance style and mode for most of the dimensions are not uncommon and so meta-governance frameworks are often constructed from inherent biases. The process of selecting styles and modes is such a natural process that those responsible for projects often use meta-governance principles without ever hearing of the term (Meuleman, 2008). This appeared to hold true as the meta-governance strategies and the governance frameworks employed in several of the following case studies also occurred without a specific governance strategy.

Table 8.1 Each governance style uses a particular mode or method to address the various dimensions of governance.

Dimensions	Modes of governance styles		
	Hierarchy	Market	Network
Orientation	Top-down	Bottom-up	Reciprocity
Role in society	Ruler	Service	Partner
Metaphor	Stick	Carrot	Brain
View of actors	Subjects	Clients	Partners
Control	Authority	Price	Trust
Strategy	Planning	Entrepreneur	Learning
Relations	Dependent	Independent	Interdependent
Communication mode	Informing	Marketing	Dialogue
Interaction type	Telling	Bargaining	Debating
Legal	Legislation	Contract	Covenant
Output	Law	Service/product	Agreement
Transaction	Unilateral	Bi-, multilateral	Multilateral

Note: Meta-governors may select particular governance styles and methods to construct meta-governance frameworks.

8.4   Case study analysis

The purpose of analyzing the 11 case studies was to understand governance actors and roles, governance styles and frameworks, and governance footprints of the agriculture sustainability efforts. This information was used to identify meta-governance strategies and any governance shifts from both an organizational perspective and as a comparison among the case studies. The 11 case studies were divided into three groupings.

The first grouping is a comparison of the two eras of the USDA CDS: the Soil Conservation Service (SCS) from 1934 to 1994 and the current era of the Natural Resources Conservation Service (NRCS) from 1994 to present. The second grouping contains three independent, but related environmental quality assurance pilot projects. They occurred in a chronological order beginning with a 2001–2009 milk industry project, a 2010–2011 state government project, and a 2013–2014 local conservation district project.

The third grouping contains six emerging, nontraditional and rather prominent regional and national efforts to account for and promote agricultural landscape sustainability. These include the efforts of an agriculture supply chain, a corporate supply chain, the federal government, a state government, a utility, and an agriculture retailer. Their goals are to develop an assessment process to account for agricultural landscape sustainability and in some cases generate and transact value for landscape sustainability.

8.4.1   Case assessment strategy

Four components and their subcomponents were assessed by identifying the governance actor types and the governance styles that led to the overall governance framework. Governance actors were identified using the governance compass sectors and governance styles were determined by assessing how decisions were made, how information was gathered, and how actions were completed. Four components and subcomponents were included:

1. Project development: Six subcomponents were identified as initiation, sources of funding, sustainability objectives, assessment process, setting standards, and accounting methods. Accounting methods were identified as (P) practice-based (identifying specific activities conducted and structures installed), (O) outcome-based (accounting methods using indices or metrics), or (P–O) a combination or hybrid method.

2. Delivery system: Five subcomponents were identified as data collection, data management, planning, implementation, and training/education. This component identifies the stakeholders responsible for the data and processes related to accomplishing the project tasks.
3. Oversight: Two subcomponents were identified as assurance and auditing. This component identifies the stakeholders responsible for the integrity of data, practices, and actions applied.
4. Valuation types and sources: P and/or O are used to describe if the value is accounted for using practice-based and/or outcome–based methods. The source of the value refers to the entity providing payment or other exchanges of value exchange. The governance style was based on whether the transaction was mandatory (hierarchy), voluntary with economic incentives (market), or voluntary with recognition benefits (network). Value types and the criteria include the following:

   a. Compliance/regulatory assurance—meets government standards
   b. Reasonable assurance—meets the project's standards
   c. Recognition/knowledge—maintains a level of sustainability
   d. Market access—meets supply chain sustainability criteria
   e. Program participation—meets sustainability criteria to enroll in a program
   f. Cost-share/reimbursement—meets program implementation standards
   g. Payment for eco-services—produces qualified and/or quantified eco-service
   h. Liability premium credit—meets management criteria for reducing risk
   i. Eco-service credit—produces a qualified and/or quantified eco-system good or service

8.4.2   Governance actors

The assessment tables (Tables 8.2 through 8.12) identified primary and participating governance actors for each subcomponent. The primary governance actor is denoted with a capitalized bold "**X**." A lowercase "x" identifies the participating governance actors. A Governance Style column rated each subcomponent as H, M, or N. These ratings were used to determine the governance styles of the four components and were used as a factor in determining the overall or predominant governance style of each project. At the bottom of the tables, Primary Actor Totals, Primary Actor %, Participation Totals, and Participation % are

Table 8.2 USDA Soil Conservation Service conservation delivery system framework of governance actors and governance styles of development, delivery, oversight, and valuation components

| Project components | Governance Actors | | | | Governance style (H/M/N) |
| | Policy makers | | Practitioners | | |
	Public	Private	Public	Private	
Development					H-6
Initiator(s) of project/program	X				H
Funding	X				H
Determine sustainability objectives	X		x		H
Assessment process/template	X		x		H
Acct. method—practice	X				H
Set standards	X		x		H
Delivery					H-4, M-1
Data collection			X		H
Data management/reporting	X		X		H
Conservation/supply chain plans			X	x	H
Implement practices	x		X	x	M
Training/education	X		x		H
Oversight					H-2
Assurance	x		X		H
Auditing	x		X		H
Valuation type(s) and source(s)					M-1
P—cost-share/reimbursement	X				M
Primary actor totals—14	8	0	6	0	H-12, M-2
Primary actor %	57	0	43	0	
Participation totals—24	12	0	10	2	
Participation %	50	0	42	8	

Note: **X** denotes the primary governance actor and x denotes the participating governance actors.

Table 8.3 USDA Natural Resources Conservation Service conservation delivery system framework of governance actors and governance styles of development, delivery, oversight, and valuation components

Project components	Policy makers Public	Policy makers Private	Practitioners Public	Practitioners Private	Governance style (H/M/N)
Development					H-6
Initiator(s) of project/ program	X				H
Funding	X				H
Determine sustainability objectives	X	x	x	x	H
Assessment process/ template	X		x		H
Acct. method—P/O	X				H
Set standards	X		x		H
Delivery					H-4, M-1
Data collection			X	x	H/m
Data management/ reporting	x		X	x	H/m
Conservation/supply chain plans	x	x	X	x	H
Implement practices	x	x	X	x	M
Training/education	x	x	X	x	H/n
Oversight					H-2
Assurance	x	x	X	x	H/m
Auditing	x		X		H
Valuation type(s) and source(s)					H-1, M-3
P—cost-share/ reimbursement	X				M
P—conservation compliance	X				H
O—payment for eco-services	X				M
O—program participation	X				M
Primary actor totals—17	10	0	7	0	H-13, M-4
Primary actor %	59	0	41	0	
Participation totals—38	16	5	10	7	
Participation %	42	13	26	18	

Note: **X** denotes the primary governance actor and x denotes the participating governance actors.

Table 8.4 MMPA Environmental Quality Assurance framework of governance actors and governance styles for development, delivery, oversight, and valuation components

Project components	Governance actors				Governance style (H/M/N)
	Policy makers		Practitioners		
	Public	Private	Public	Private	
Development					H-2, N-4
Initiator(s) of project/program		X			H
Funding	X	x			H
Determine sustainability objective	x	X	x	x	N-l
Assessment process/template	x	X	x	x	N-l
Acct. method—practice	x	X	x	x	N-l
Set standards	x	X	x	x	N-l
Delivery					H-1, M-3, N-1
Data collection			x	X	M
Data management/ reporting		X	x	x	H
Conservation/ supply chain plans			x	X	M
Implement practices			x	X	M
Training/education	x	X	x	x	N-l
Oversight					H-2
Assurance		X			H
Auditing		X			H
Valuation type(s) and source(s)					M-2
P—cost-share/ reimbursement		X			M
P—five star certification		X			M
Primary actor totals—15	1	11	0	3	H-5, M-5, N-5
Primary actor %	7	73	0	20	
Participation totals—36	6	12	9	9	
Participation %	17	33	25	25	

Note: **X** denotes the primary governance actor and x denotes the participating governance actors.

Table 8.5 MN Dept. of Agriculture Livestock Environmental Quality Assurance framework of governance actors and governance styles for development, delivery, oversight, and valuation components

Project components	Governance actors				Governance style (H/M/N)
	Policy makers		Practitioners		
	Public	Private	Public	Private	
Development					H-2, N-4
Initiator(s) of project/program	X				H
Funding	X				H
Determine sustainability objectives	X	x	x	x	N-a
Assessment process/template	x	X	x	x	N-a
Acct. methods—P/O	x	X	x	x	N-a
Set standards	x	X	x	x	N-a
Delivery					H-1, M-3, N-1
Data collection			x	X	M
Data management/reporting	x		x	X	H
Conservation/supply chain plans	x		x	X	M
Implement practices	x		x	X	M
Training/education	X	x	x	x	N-a
Oversight					H-2
Assurance	x		X	x	H
Auditing	X			x	H
Valuation type(s) and source(s)					M-3
P—cost-share/reimbursement	X				M
P—reasonable assurance	X				M
O—reasonable assurance	X				M
Primary actor totals—16	8	3	1	4	H-5, M-6, N-5
Primary actor %	50	19	6	25	
Participation totals—41	15	5	10	11	
Participation %	37	12	24	27	

Note: **X** denotes the primary governance actor and x denotes the participating governance actors.

Table 8.6 Sunrise AgEQA Watershed Project framework of governance actors and governance styles for development, delivery, oversight, and valuation components

| Project components | Governance actors | | | | Governance style (H/M/N) |
| | Policy makers | | Practitioners | | |
	Public	Private	Public	Private	
Development					H-1, N-5
Initiator(s) of project/program	x		**X**		N-l
Funding	**X**		x		H
Determine sustainability objectives			**X**	x	N-p
Assessment process/ template			**X**	x	N-p
Acct. method—outcome	x		**X**	x	N-p
Set standards			**X**	x	N-p
Delivery					H-1, M-3, N-1
Data collection			x	**X**	M
Data management/ reporting	x		**X**	x	H
Conservation/supply chain plans			x	**X**	M
Implement practices			x	**X**	M
Training/education			x	**X**	N-a
Oversight					H-2
Assurance			**X**	x	H
Auditing			**X**	x	H
Valuation type(s) and source(s)					M-1
O—voluntary resource objective			**X**		M
Primary actor totals—14	1	0	9	4	H-4, M-4, N-6
Primary actor %	7	0	64	29	
Participation totals—29	4	0	14	11	
Participation %	14	0	48	38	

Note: **X** denotes the primary governance actor and x denotes the participating governance actors.

Table 8.7 MN Agriculture Water Quality Certainty Program framework of governance actors and governance styles for development, delivery, oversight, and valuation components

| | Governance actors | | | | Governance style (H/M/N) |
| | Policy makers | | Practitioners | | |
Project components	Public	Private	Public	Private	
Development					H-2, N-4
Initiator(s) of project/ program	X				H
Funding	X				H
Determine sustainability objectives	X	x	x	x	N-l
Assessment process/ template	X	x	x	x	N-l
Acct. method— practice/outcome	X	x	x	x	N-l
Set standards	X	x	x	x	N-l
Delivery					H-4, M-1
Data collection	X		**X**	x	H
Data management/ reporting	X		x	x	H
Conservation/supply chain plans	X		x	x	H
Implement practices	X		**X**	x	M
Training/education	X		x		H
Oversight					H-2
Assurance	X		x		H
Auditing	X				H
Valuation type(s) and source(s)					M-4
P—cost-share/ reimbursement	X				M
P—regulatory certainty	X				M
O—regulatory certainty	X				M
O—payment for eco-services		**X**			M
Primary actor totals—17	14	1	2	0	H-8, M-5, N-4
Primary actor %	82	6	12	0	
Participation totals—39	16	5	10	8	
Participation %	41	13	26	20	

Note: **X** denotes the primary governance actor and x denotes the participating governance actors.

Table 8.8 Chesapeake Bay Program framework of governance actors and governance styles for development, delivery, oversight, and valuation components

Project components	Governance actors				Governance style (H/M/N)
	Policy makers		Practitioners		
	Public	Private	Public	Private	
Development					H-2, N-4
Initiator(s) of project/program	X				H
Funding	X				H
Determine sustainability objectives	X	x	x	x	N-l
Assessment process/template	X	x	x	x	N-l
Acct. method—practice	X	x	x	x	N-l
Set standards	X	x	x	x	N-l
Delivery					H-2, M-2, N-1
Data collection	x	x	X	x	H
Data management/ reporting	X		x	x	H
Conservation/ supply chain plans	x	x	X	x	M
Implement practices			X	x	M
Training/education	X	x	x	x	N-l
Oversight					H-2
Assurance	x	x	X	x	H
Auditing	x		X	x	H
Valuation type(s) and source(s)					
N/A					
Primary actor totals—13	8	0	5	0	H-6, M-2, N-5
Primary actor %	62	0	38	0	
Participation totals—42	12	8	11	11	
Participation %	29	19	26	26	

Note: **X** denotes the primary governance actor and x denotes the participating governance actors.

Table 8.9 Electric Power Research Institute Ohio River Basin water quality trading framework of governance actors and governance styles for development, delivery, oversight, and valuation components

| Project components | Governance actors | | | | Governance styles (H/M/N) |
| | Policy makers | | Practitioners | | |
	Public	Private	Public	Private	
Development					N-6
Initiator(s) of project/program	x	**X**	x	x	N-1
Funding	x	**X**	x	x	N-1
Determine sustainability objectives	x	**X**	x	x	N-1
Assessment process/template	x	**X**	x	x	N-1
Acct. method—practice	x	**X**	x	x	N-1
Set standards	x	**X**	x	x	N-1
Delivery					H-2, M-2, N-1
Data collection		x	**X**	x	H
Data management/ reporting		**X**	x	x	H
Landscape/supply chain plans			**X**	x	M
Implement practices			**X**	x	M
Training/ education	x	**X**	x	x	N-1
Oversight					H-2
Assurance		**X**	x	x	H
Auditing		**X**	x	x	H
Valuation type and source					M-1
Water quality credits		**X**			M
Primary actor totals—14	0	11	3	0	H-4, M-3, N-7
Primary actor %	0	79	21	0	
Participation totals—45	7	12	13	13	
Participatory %	16	26	29	29	

Note: **X** denotes the primary governance actor and x denotes the participating governance actors.

Table 8.10 The Sustainability Consortium framework of governance actors and governance styles for development, delivery, oversight, and valuation components

Project components	Governance actors				Governance style (H/M/N)
	Policy makers		Practitioners		
	Public	Private	Public	Private	
Development					M-1, N-5
Initiator(s) of project/program	X	**X**	x	x	N-l
Funding	X	**X**	x	x	M
Determine sustainability objectives	X	**X**	x	x	N-a
Assessment process/template	x	**X**	x	x	N-a
Acct. method—P/O	x	**X**	x	x	N-a
Set standards	x	**X**	x	x	N-a
Delivery					M-4, N-1
Data collection		x		**X**	M
Data management/ reporting		**X**		x	M
Conservation/ supply chain plans		**X**		x	M
Implement practices		x		**X**	M
Training/education	x	**X**	x	x	N-l
Oversight					N-2
Assurance		x		**X**	N-p
Auditing		x		**X**	N-p
Valuation type					M-1
P/O—market access		**X**			M
Primary actor totals—14	0	10	0	4	M-6, N-8
Primary actor %	0	71	0	29	
Participation totals—41	7	14	7	13	
Participation %	17	34	17	32	

Note: **X** denotes the primary governance actor and x denotes the participating governance actors.

Table 8.11 Field to Market framework of governance actors and governance styles for delivery, development, oversight, and valuation components

| Project components | Governance actors | | | | Governance style (H/M/N) |
| | Policy makers | | Practitioners | | |
	Public	Private	Public	Private	
Development					N-6
Initiator(s) of project/program		**X**			N-1
Funding	x	**X**	x	x	N-1
Determine sustainability objectives	x	**X**	x	x	N-a
Assessment process/template	x	**X**	x	x	N-a
Acct. method—outcome	x	**X**	x	x	N-a
Set standards	x	**X**	x	x	N-a
Delivery					M-4, N-1
Data collection				**X**	M
Data management/ reporting		x		**X**	M
Conservation/ supply chain plans			x	**X**	M
Implement practices			x	**X**	M
Training/education	x	**X**	x	x	N-a
Oversight					M-2
Assurance		x		**X**	M
Auditing		x		**X**	M
Valuation type					M-1
O—voluntary resource objectives		**X**			M
Primary actor totals—14	0	8	0	6	M-7, N-7
Primary actor %	0	57	0	43	
Participation totals—37	6	11	8	12	
Participation %	16	30	22	32	

Note: **X** denotes the primary governance actor and x denotes the participating governance actors.

Table 8.12 United Suppliers, Inc. SUSTAIN framework of governance actors and governance styles for development, delivery, oversight, and valuation components

Project components	Governance actors				Governance style (H/M/N)
	Policy makers		Practitioners		
	Public	Private	Public	Private	
Development					N-6
Initiator(s) of project/program		x		X	N-1
Funding	x	x		X	N-1
Determine sustainability objectives	x	x		X	N-1
Assessment process/template		x		X	N-1
Acct. method— practice/metrics		x		X	N-1
Set standards		x		X	N-1
Delivery					H-3, M-2
Data collection		x		X	H
Data management/ reporting		x		X	H
Conservation/ supply chain plans				X	M
Implement practices				X	M
Training/education		x		X	H
Oversight					H-2
Assurance		x		X	H
Auditing		x		X	H
Valuation type(s) and sources(s)					M-1
P/O—voluntary resource objectives		X		x	M
Primary actor totals—14	0	1	0	13	H-5, M-3, N-6
Primary actor %	0	7	0	93	
Participation totals—28	2	12	0	14	
Participation %	7	43	0	50	

Note: **X** denotes the primary governance actor and x denotes the participating governance actors.

provided. Identification of the governance styles of subcomponents, components, and the case studies is somewhat subjective as each may contain mixtures of governance styles that cannot be objectively delineated.

8.4.3   Governance styles, frameworks, and footprints

Governance styles for each task were noted as H, M, N-l, N-a, and N-p for hierarchy, market, network-lead, network-administrative, and network-participant, respectively. In review, *hierarchy* is a top-down conformist structure, *market* is a top-down profit structure, *network-lead* is the use of an internal entity to administrate and facilitate a network, *network-administrative* is the use of an external entity to administrate and facilitate network, and *network-participant* is a self-organized structure which participants engage as they deem necessary.

The overall project governance framework consists of actors and styles used for the project components and subcomponents. A *governance footprint* is a representation of the ratio of governance styles used to achieve project objectives (Meuleman, 2008). Governance footprints, for these case studies, were plotted by the predominant governance style of the framework and the method used to account for eco-services.

8.5   Group I case studies: USDA conservation delivery system

The two eras of the USDA CDS, the SCS from 1934 to 1994 and the NRCS from 1994 to present, were chosen as a comparative case study because of the USDA's historical prominence and its continual influence with agriculture landscape decisions. The SCS was the foundation for the nation's early conservation movement and the NRCS continues to influence how corporate sustainable supply chains, water quality trading programs, and the emerging ecosystem service markets are structured. From a governance perspective, the two eras provide a unique opportunity to compare how accounting strategies and governance styles evolved within a single organization as the agricultural production system became increasingly complex.

8.5.1   Soil Conservation Service CDS

The idea for the SCS was born from the recognition that soil erosion negatively affects the agricultural landscape from economic and ecological perspectives. Hugh Bennett, considered the father of American conservation, was profoundly affected by a land survey in Louisa

County, Virginia he conducted in 1905. He had been directed to investigate declining crop yields and as he compared virgin, timbered sites to eroded fields, he became convinced that soil erosion was a problem not just for the individual farmer, but also for rural economies (Helms, 2009).

In November 1934, Bennett recommended that states should be encouraged to pass legislation authorizing (1) cooperation with the federal government in erosion control; (2) the organization of conservation districts to carry out measures of erosion control; and (3) the establishment of state or local land-use zoning ordinances where lack of voluntary cooperation make such ordinances necessary. The legislation would not be just an experiment in conserving natural resources, it would be an experiment in the application of government [and governance] to natural-resource problems in a democracy (Sampson, 1985).

8.5.1.1   Project components

On April 27, 1935, the 74th Congress enacted Public Law No. 46 to provide for the protection of land resources against the wastage of soil and moisture resources on farm, grazing, and forest lands. The first soil conservation district was chartered August 4, 1937, and by 1950, the number of districts climbed to 2,164. By this time, legislation enabling state agencies was developed and the progression of the organization of conservation districts led to the National Association on Conservation Districts and the now-called "conservation delivery system" was in place (Sampson, 1985).

A three-phase planning process was the delivery method for technical and financial assistance. SCS and conservation district staff met with farmers on-site to (1) assess their resource needs, (2) formulate and discuss options, and (3) implement and evaluate the plan. Oversight was conducted by SCS and/or conservation district staff to ensure conservation practices were installed and were functioning properly according to the USDA Field Office Technical Guide. A certain percentage of practices were audited for design and installation integrity, and field verified. Farmers were reimbursed by the USDA for a portion of the cost of the practice.

8.5.1.2   Assessment and discussion

The development, delivery, and oversight components of the SCS conservation delivery system were based on a hierarchy governance style with 100% of the primary actors being public-based and the majority (92%) of decisions and activities carried out by public policy makers and public practitioners (Table 8.2). The primary actors were public policy makers (8) and public practitioners (6). Farmers, the private practitioners, decided which practices would best meet their needs within the context of the programs and SCS approval authority. The valuation component was an incentive (market governance style) cost-share payment from the USDA to the farmer.

By many accounts, this predominantly hierarchical system was considered to be extremely successful in improving the soil-resource base of agriculture and the farming economy (Phillips, 2007). In hindsight, the success of the SCS may have been its near singular control on policy and programs, a singular focus on the soil resource, and the relatively evenly dispersed, national, skilled workforce employed to address the concerns of individual farmers. The hierarchical organizational structure was sufficient to manage the relatively low level of complexity inherent to early and mid-century American agriculture.

8.5.2   Natural Resources Conservation Service CDS

The second era of the conservation delivery system officially began in 1994 when the SCS was renamed the NRCS to represent its broader objectives in conservation that began in the decade prior (103rd Congress, 1994). In the 1985 Farm Bill, conservation compliance and the Conservation Reserve Program (CRP) were written to law. CRP provided payments to producers to put environmentally sensitive cropland into conservation uses for 10–15 years. Conservation compliance required farmers to manage their highly erodable lands (HEL) to reduce soil erosion and placed restrictions on draining wetlands. The 1990 Farm Bill created a federal program to restore and place conservation easements on wetlands signaling the emergence of water quality as a primary environmental objective of conservation programs.

The NRCS vastly increased funding for conservation on lands in crop and animal production, or "working lands," by authorizing increased spending through the Environmental Quality Incentives Program (EQIP) and Conservation Stewardship Program (CSP): a program to provide incentives for environmental *outcomes* on working lands. In the 2014 Farm Bill, the Regional Conservation Partnership Program (RCPP) was created to encourage partners to join in efforts with producers to increase the restoration and sustainable use of soil, water, wildlife, and related natural resources on regional or watershed scales (NRCS, 2015).

8.5.2.1   Project components

In 1994, Congress initiated a major reorganization and renamed the SCS the NRCS to better reflect the broad scope of the agency's mission; this marked the beginning of the NRCS's growing responsibility for administering financial assistance for conservation programs (USDA NRCS, 2015).

The NRCS still relies on the three-phase planning process to deliver technical and financial support to farmers. The USDA uses a winnowing process (Schnepf and Cox, 2007) to choose enrollees. This process begins with a sign-up process to determine which producers, land, and practices are eligible. The second step is producer application where the producer decides what land, practices, and perhaps, the payment amount she/he is willing to accept in exchange for the conservation practices. In the third and final step, the agency decides which bids or applications to accept or reject.

Farmers receive assistance in this process from the NRCS, local conservation district staff, and technical service providers (TSPs). TSPs are individuals or organizations from the public or private sector that have technical expertise in conservation activities. They are hired by farmers, ranchers, private businesses, nonprofit organizations, or public agencies to provide these services on behalf of the NRCS. Each certified TSP is listed on the NRCS TSP online registry, TechReg. The TSP registration and approval process involves required training and verification of essential education, knowledge, skills, and abilities (USDA TSP, 2015).

Oversight is conducted by the NRCS, conservation district staff, and/or TSPs to ensure conservation practices and activities are installed and functioning properly according to the USDA Field Office Technical Guide. For approved conservation practices, conservation costs are shared between the farmers, NRCS, and in some cases, other government agencies, corporations, and NGOs. The USDA also reinstituted conservation compliance requirements for farmers to participate in government-subsidized crop insurance programs, a mandatory practice for those participating in USDA programs. As the NRCS's first "payment for ecosystem services" program, the CSP uses land management indices to determine eligibility and payment levels. Farmers enroll in a multiyear contract and are provided payment based on environmental outcomes and specific practices.

8.5.2.2 *Assessment and discussion*

The NRCS development, delivery, and oversight components are each based on a hierarchy governance style. The lead governance actors are all public-based and the majority (68%) of the participation (decision and activities) is carried out by public policy makers and public practitioners (Table 8.3). Corporations and NGOs (private policy makers) have a slight, but increasing influence on how objectives are determined and delivered.

The governance of the delivery component has shifted slightly toward market governance (denoted with /m) with the NRCS adopting the TechReg Program that incorporates private practitioners (agriculture and conservation professionals) into the delivery component. A 2005 report stated that of the $53 million going to the TechReg Program, 62% went to the private sector, and the remaining to conservation districts (USDA, 2006). The NRCS also engages with numerous local, state, and federal government, NGOs, and the private sector and is expanding its network governance with the Regional Conservation Partnership Program in 2014. Oversight is predominantly hierarchical, with some TSPs providing assurance on practice completion.

The valuation component includes outcome and practice-based systems administered by the NRCS. Participation is based on incentive-based market governance for all programs with the exception of the mandatory conservation compliance for those who enroll in any USDA program.

8.5.3 Group I discussion

In both eras, the public sectors are the primary governance actors as Figure 8.3 shows nearly identical governance sector ratios. (The slight increase in public policy-maker percentage for the NRCS is due to the additional valuation types.)

A more noticeable shift occurred between the two eras relative to governance actor participation (Figure 8.4). With the adoption of TechReg, the NRCS shifted toward market governance in the delivery component by contracting with public and private sector professionals for technical assistance. Coincidently, the NRCS also adopted outcome-based indices to determine program eligibility and payments for the CRP and CSP programs. This shift caused the number of governance actors participating in the SCS era (24) to increase in the NRCS era (38), with the increase coming from the private policy maker and practitioner sectors.

Heywood (2003) noted the addition of private practitioners and this type of governance shift is characteristic of new modes of delivering public services. It reflects a move away from direct service provision to an

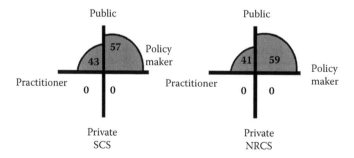

Figure 8.3 Primary governance actors for the conservation delivery system of both the SCS and the NRCS reside in the public sector and remained relatively the same between the two eras.

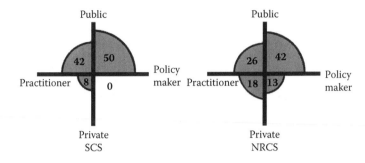

Figure 8.4 Total governance actor participants increased in the NRCS era as private practitioners and policy makers became actively engaged in the CDS. The percentage of public sector participants shifted from 92% in the SCS CDS down to 68% in the NRCS CDS. This shift in governance actors was due to the TechReg incorporating private sector professionals to conduct work traditionally handled by public practitioners.

enabling role and the increased use of private organizations to deliver public services.

Avril and Zumello (2013) noted that as organizations evolve to meet new challenges, their new governance styles are characterized by increased collaboration among private, public, and nonprofit organizations. This reflects that, as society has become more complex and fluid, new methods of governing have had to be devised. These new governance strategies rely less on hierarchical state institutions and more on market and network governance, thus blurring the distinction between the state and society. As the NRCS shifted its focus from soil resources to broader environmental objectives, it responded by expanding its technical support base. These additional professionals were not brought in-house within a hierarchy governance structure, but are external partners engaging with the conservation delivery system within a market governance model.

The overall governance shifted slightly from an SCS ratio of H-12, M-2, and N-0 (Table 8.2) to an NRCS ratio of H-13, M-4, and N-0 (Table 8.3).

8.6   *Group II case studies: Minnesota EQAs*

The second group includes three separate, but related Minnesota (USA)-based projects involving on-farm assessment and environmental quality assurance objectives. These studies illustrate a continuum of sorts, of shifting governance styles and evolving accounting strategies. In 2001, the Minnesota Milk Producers Association (MMPA) developed an Environmental Quality Assurance Program (MMEQA) to develop a more inclusive and streamlined conservation delivery system and to account for agriculture sustainability. Its success led to the Minnesota legislature supporting a Livestock Environmental Quality Assurance Program (LEQA), a state agency administered program from 2010 to 2011. In 2013–2014, a local conservation agency applied this concept as an Agriculture Environmental Quality Assurance (AgEQA) project with the objective of creating a more effective resource assessment process and accounting for watershed improvements.

8.6.1   Minnesota Milk Producers Association EQA

The MMEQA (2001–2009) pilot project was initiated as an on-farm assessment process to create a more responsive and integrated conservation delivery system and to account for environmental benefits. It represented an early industry-developed environmental assurance process to address the dual issues of emerging sustainability demands of corporations and a more stringent feedlot enforcement approach by the Minnesota Pollution Control Agency (MPCA).

8.6.1.1   Project components
The MMEQA project was initiated by the MMPA and funded by the Legislative Citizen Commission of Minnesota Resources (LCCMR). Multiple partners, including local, state, and federal government staff, NGOs, and agribusinesses provided input to develop the program with leadership provided by the MMPA.

The delivery component relied on private and public practitioners to conduct the on-farm assessments on 100 farms. MMEQA technicians were hired by dairy producers to complete the assessment and assurance process. Training and certification was designed and delivered under supervision of the MMPA. Oversight and auditing was administered by the MMPA. A practice-based accounting method was used to calculate the MMEQA assessment "score." A farm operation achieving the MMEQA standards was recognized by the MMPA as achieving a five-star rating for sustainability.

8.6.1.2   Assessment and discussion
A network-lead (N-l) governance style was the primary governance style used in the development component (Table 8.4). In a network-lead structure, an entity (MMPA) internal to the network takes on the role of administration and facilitation of the effort. In network-lead governance, all major network-level activities and key decisions are coordinated through and by a single participating member and may result in centralized governance with power residing with the lead organization (Provan and Kenis, 2007). In these cases, network-lead governance may have characteristics of hierarchy governance, in that a single entity provides administration and oversight, but the process is inclusive and the decisions are often made by consensus.

The delivery component were predominantly market governance as decisions to provide technical assistance were based on compensation. Oversight and auditing were conducted under the supervision of the MMPA using a top-down hierarchy governance protocol. Values associated with the five-star certification and reimbursement for applying practices were provided through the MMPA with decisions based on incentives (market governance).

The MMEQA applied the three governance styles (H-5, M-5, and N-5) in equal proportions; this is an example of a balanced meta-governance strategy, where a mix of governance styles is used for specific tasks and objectives. The application of specific governance styles was not necessarily a conscious decision by the MMPA, but more of a natural progression in seeking out the most favorable delivery method for their dairy farmer members and the project participants. MMPA, as a private policy-making organization, was the primary actor (12) with a total of 36 governance actor types participating in the four project components.

8.6.2 Minnesota Department of Agriculture's LEQA

The second case study in Group II is a Minnesota Department of Agriculture's (MDA) LEQA project from 2010 to 2011. The 2009 Minnesota Legislature directed the MDA to develop a water quality improvement program with the capacity to provide *reasonable assurance* of the water quality effects (MN Leg, 2010). Its format was based on the MMEQA and it was expanded to include all livestock operation types.

8.6.2.1 Project components

Program development began with an issuance of a request for proposal (RFP) by the MDA with the contract awarded to Ag Resource Strategies, LLC (AgRS), a private business. The Livestock Environmental Assurance Consortium (LEAC) was the primary stakeholder group providing guidance and input in program development. The LEAC consisted primarily of nonprofit farm organizations involved with grain and livestock production with most of them involved in some aspect of the MMEQA.

In consultation with the LEAC, a LEQA assessment template was created based on the MMEQA format. The two major changes were using farm management units rather than natural resource parameters for the assessment categories and the use of land management indices to account for soil, habitat and water quality parameters rather than relying solely on practice-based accounting. These changes created *landscape intelligence*, a scalable geographically based data set consisting of an index value per land area.

The delivery component relied on certified LEQA technicians from public and private sectors. Assurance and auditing was provided by AgRS within the parameters determined by the LEAC and MDA.

The valuation component consisted of a reimbursement for a portion of the costs of installing conservation practices and the five-star certification, an industry recognition and the potential for a *reasonable assurance* recognition of achieving state water quality objectives.

8.6.2.2 Assessment and discussion

The MDA LEQA applied the three governance styles (H-5, M-6, and N-5) in fairly equal proportions as a mixed or meta-governance strategy (Table 8.5). A network-administrative (N-a) governance style was considered the predominant governance style in the development component. AgRS, an external entity, provided the bulk of the administrative and facilitation of stakeholders. Market governance was the predominant style for program delivery with the use of public and private practitioners choosing to participate based on compensation.

A top-down hierarchy process was used for assurance and auditing protocol as dictated by MDA with input from the LEAC. Valuation types included cost-share incentives and a *reasonable assurance* value. Reasonable assurance was intended to provide farmers with a means to account for their land management relative to state water quality goals. The assurance template was a combination of practice-based and outcome-based methods. The source of the cost-share incentive and reasonable assurance was state government.

8.6.3 Chisago SWCD's AgEQA

The Sunrise AgEQA Watershed Project (AgEQA) was initiated by the MPCA and the Chisago Soil and Water Conservation District (SWCD) with the purpose of increasing technical assistance capacity and to account for natural resource benefits within the Sunrise Watershed. A third objective was to develop an accounting and valuation strategy for water quality within the context of meeting the state's total maximum daily load (TMDL) limits.

An index-based asset portfolio, containing data similar to Table 7.1, was developed to account for outcomes based on farm management units (fields, livestock facilities, farmstead, and overall farmscape). Each resource asset was scored on a 0–100 scale.

The portfolio was to serve several purposes. First, it served as a baseline to account for a farm's resource assets and a sustainability *starting line* for the farmers. Second, it described the resource objective, or the *finish line* that farmers needed to meet for a particular resource. Third, it served as an adaptive management platform for engagement of agriculture and conservation professionals in achieving resource outcomes. Fourth, it served as a potential valuation platform for external entities such as state and federal governments, NGOs, corporations, and watershed organizations to express the management level desired or demanded of certain resource assets and the outcomes. The index-based portfolio had several of the *smart* assessment attributes described in Chapter 7.

8.6.3.1   Project components

The AgEQA was funded by the MPCA and administered by the Chisago SWCD. A private business, AgRS, was contracted to develop assessment templates, deliver training, and facilitate interaction among stakeholders. Agriculture production expertise was provided by a local agronomist. Final decisions on sustainability objectives, assessment templates, accounting methods, and standards were determined by consensus of these primary stakeholders.

The delivery component relied on a mix of stakeholders. AgRS provided training to SWCD staff and agronomists. Field data was collected by farmers and their advisors on six farms. Indices were calculated by the advisor or the SWCD staff depending on the preference of the farmer. Data was compiled and reviewed by AgRS. Assurance and auditing tasks were completed by the SWCD. The valuation component used an outcome-based index to seek potential values from various entities, such as state government or corporate supply chain managers.

8.6.3.2   Assessment and discussion

The AgEQA applied the three governance styles (H-4, M-4, and N-6) in a fairly equal ratio as a mixed or meta-governance strategy (Table 8.6). A network-participant (N-p) governance style was considered the predominant style in the development component. The SWCD provided the administration, AgRS provided the facilitation and training, and all the participants were involved in deciding what type of indices should be used and setting standards. These roles were determined jointly by the group with each favoring their expertise.

The delivery component used market governance. The tasks and compensation rates were publicized to the community and individuals voluntarily participated based on their business and economic objectives. The assurance and auditing tasks were conducted by the SWCD by using a prescribed top-down process. The valuation type was the recognition by the SWCD that the farm operation met the AgEQA standards and a future potential of value. The AgEQA portfolio approach identified on-farm natural resources as ecological assets that could be improved, rather than environmental liabilities that needed to be corrected. These assets represented potential value to the broad range of entities interested in achieving agricultural landscape sustainability.

8.6.4   Group II discussion

The chronological progression of the three pilot projects provided a unique opportunity to study the application and evolution of accounting methods and governance styles within similar contexts. The network-lead and practice-based accounting used by the MMEQA allowed it to control the process, as there was much uncertainty in this relatively new venture. Network-lead often acts as hierarchy governance, in that an internal entity administers and facilitates the process. The MMEQA was as relatively early sustainability project (2001) and its structure was based on the well-known and accepted practice-based system of the NRCS.

The LEQA project was a direct offshoot from the MMEQA. As a state legislative effort, it was prone to politicization and a network-administrative model helped diffuse some of those issues by contracting with an external entity. The LEQA accounting method was expanded to include landscape indices to reduce the data management costs and potentially increase its value in addressing other stakeholders' sustainability data demands. The potential value of the index-based template emerged when a cheese processor procuring milk and an electric utility using crop biomass sought a means to account for landscape sustainability. Both entities were seeking an accounting and valuation system to address sustainability claims and demands within their industry. This additional valuation process never materialized, but their requests highlighted the need for a streamlined and multiuse assurance process that could simultaneously address the similar sustainability demands from disparate stakeholders. It introduced the concept of shared governance to agriculture landscape sustainability efforts.

The AgEQA was based on the findings of these two previous pilot projects with the goal that watershed sustainability needed a more inclusive approach and streamlined data process. What resulted was a network-participant governance style that shared the decision making, verification, and data management tasks among the primary stakeholders. The AgEQA effort combined the community connections and experience of professional agronomists to compile data and compute indices. The indices were scaled from 0 to 100, creating uniformity on how resource management was accounted for across resource types.

The governance compasses (Figure 8.5) illustrate how the three case studies varied relative to the primary governance sectors. The MMEQA was directed primarily by private policy makers with a significant shift occurring to public policy makers as the legislators assigned the LEQA program to the MDA. The AgEQA, as a locally driven project, was directed primarily from the perspective of local public practitioners.

Despite the relatively different primary governance actors for each Group II study, the ratios of governance participants are fairly well-rounded as shown by the governance compasses in Figure 8.6.

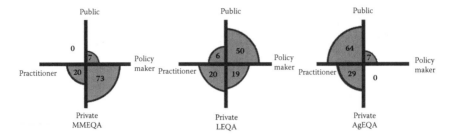

Figure 8.5 The primary governance actors for the MMEQA, LEQA, and AgEQA projects differed significantly depending on the source of leadership and funding. MMPA, a nonprofit organization, initiated the MMEQA industry-led project. The Minnesota Department of Agriculture along with state legislators initiated the LEQA and the Chisago SWCD, and a local government unit initiated the AgEQA.

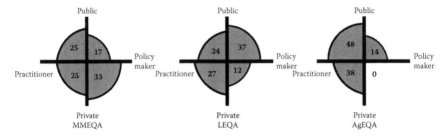

Figure 8.6 The four sectors of governance actors were well represented in the MMEQA and LEQA project with the private policy maker absent from the AgEQA. The differences may be due to the MMEQA needing wide support among its peer industry groups and government agencies. The LEQA sought political support and the AgEQA was delivered in a very localized area dependent on technical knowledge within the agricultural sector. The MMEQA and LEQA were state-wide efforts and the AgEQA was within a single county.

The three Group III case studies illustrate a progression of governance styles and accounting strategies toward more inclusive governance and outcome-based accounting.

8.7   Group III case studies: emerging strategies

The case studies in the third group represent emerging efforts focused on creating longer-term accounting and valuation sustainability platforms. Each of the six projects represents unique governance, accounting, and valuation strategies for sustaining agricultural landscapes:

- The Ag Water Quality Certainty Program (AWQCP) is a federal/ state government project addressing pending water regulations and corporate sustainability supply chains

- The Chesapeake Bay BMP Verification Project (CBay) is a federal/state project addressing TMDL regulations and goals
- The Sustainability Consortium (TSC) is a multicorporate effort addressing supply chain sustainability
- Field to Market® (FtoM) is an NGO effort addressing agriculture supply chain sustainability
- United Suppliers, Inc. (USI) is an agricultural wholesaler addressing on-farm agriculture sustainability in response to supply chain demands
- The Electric Power Research Institute, Inc. (EPRI) is an NGO addressing watershed sustainability through water quality market trading

These efforts are at early stages of development relative to their long-term goals. They were chosen as case studies due to their prominence and precedence setting potential. These case studies were also discussed in Chapter 3 in relation to describing disparate stakeholder strategy values.

8.7.1 MDA's Ag Water Quality Certainty Program

The Minnesota Agriculture Water Quality Certainty Program (AWQCP) was initiated with an agreement signed by EPA Administrator Lisa Jackson, USDA Secretary Tom Vilsack, and Minnesota Governor Mark Dayton in January 2012 (State of MN, 2012). The intention of the project is to certify producers for regulatory certainty, which means AWQCP-certified producers are deemed to be in compliance with any current or new state water quality rules or laws for a period of 10 years.

8.7.1.1 Project components

The MDA convened a broad spectrum of agriculture and environmental stakeholders to develop the program. Funds were received from the USDA NRCS and the Minnesota Legislature. Stakeholders, using a consensus process, chose a water quality assessment method from three options: the MDA LEQA model, a USDA conservation program assessment tool, and an NRCS water quality index (WQI). The AWQCP group selected the WQI and modified it for Minnesota agriculture. The four-watershed pilot project used practice- and outcome-based accounting methods. Local conservation districts conducted on-farm assessments and calculated WQI scores. Agricultural producers with a score of 8.5 or greater on a scale of 0–10 were eligible to sign a *regulatory certainty* contract. If a lessor score is received, the producer may decide to install conservation practices to increase their score. The MDA oversees training, data collection, assurance, and auditing.

The valuation component relies on both practices and the WQI score. Reimbursements for the cost of installing conservation practices are provided through the AWQCP. Farmers are also eligible to receive payments for achieving corporate supply chain objectives (Redlin et al., 2015).

8.7.1.2   Assessment and discussion

The AWQCP relies on the three governance styles (H-8, M-5, and N-4) but with an emphasis on hierarchy governance (Table 8.7). The project was initiated by a federal–state agreement and used a network-lead (N-l) governance style to develop the assessment and accounting methods, and to set standards. The network-lead governance style is more inclusive than a hierarchy model, but can function in a similar manner. The MDA acted as the network-lead.

The delivery component was developed around the WQI index and its supporting software. Field data collection and WQI calculations were assigned to local conservation district staff. Conservation plans were included with the WQI software and were a component of the contract. Training was provided by the MDA. Producers voluntarily participated in the implementation of practices based on their perceived needs and benefits (market governance). Oversight was provided by the MDA using a hierarchy governance style. A hierarchy governance style was predominant in the delivery component.

Practice- and outcome-based methods were used for valuation accounting. Valuation types included cost-share reimbursement, regulatory certainty (the primary program value), and payment for eco-services provided by a corporate food processor. These valuation types were delivered using market governance as producers choose to participate with the MDA and/or the corporate food processor based on incentives and compensation.

8.7.2   Chesapeake Bay Program BMP Verification

The Chesapeake Bay Protection and Restoration Executive Order—Executive Order 13508, signed by President Obama on May 12, 2009, called for the development of a system of accountability for tracking and reporting conservation. The Executive Order describes the full accounting of conservation practices applied to the land as "a necessary data input for improving the quality of information and ensuring that the practices are properly credited in the Bay model."

Under the 2010 Chesapeake Bay TMDL, the seven watershed jurisdictions (DE, MD, NY, PA, WV, VA, and DC) are expected to account for water pollution loadings of nutrient and sediment (U.S. EPA, 2010a). Verification, tracking, and accountability are among the elements to be addressed to ensure BMPs continue to produce the expected pollutant load reductions. The EPA was charged with developing a Chesapeake Bay BMP Verification process (CPB, 2014).

8.7.2.1 Project components

The Chesapeake Bay BMP Verification (CBay) effort was initiated and funded by the federal government and administered and facilitated by the EPA. The Chesapeake Bay Program partnership developed the Chesapeake Bay Basin–wide BMP Verification Framework, a structure to account for the BMPs applied.

This framework is to be used by stakeholders in local, regional, state, and federal agencies, institutions, organizations, and businesses involved in the implementation, tracking, verification, and reporting of BMPs for nutrient and sediment pollutant load reduction crediting. The BMPs will be reported through the Bay Program's National Environmental Information Exchange Network (NEIEN) by designated agencies. Design and implementing conservation practices are to follow NRCS guidelines. Oversight of the data is the responsibility of agencies of the seven jurisdictions. The Chesapeake Bay Program does not provide specific values for practice implementation or reporting, but land managers are potentially eligible for USDA NRCS incentive programs.

8.7.2.2 Assessment and discussion

The CBay relies primarily on a hierarchy governance style (H-6, M-2, and N-5). Network-lead governance was used in the development component with the EPA acting as lead with state governments providing leadership in their jurisdictions. The network-lead is more inclusive for stakeholder input, yet may have hierarchy governance characteristics (Table 8.8).

The delivery component is primarily a data collection system relying on government jurisdictions to account for BMPs and to process and transfer the data into the NEIEN database. Implementing conservation practices is based on a market governance model where land managers decide to implement practices based on their perceived needs and benefits. Oversight to ensure data is accurate is conducted by each jurisdiction with protocol and statistical methods developed by each jurisdiction. The CBay has no direct valuation component to land managers to implement BMPs.

8.7.3 EPRI's Ohio River Basin water quality trading

The Electric Power Research Institute, Inc. (EPRI) is an independent, non-profit 501(c)3 organization involved in a collaborative effort to improve water quality in the Ohio River Basin through implementing an interstate water quality trading program (EPRI, 2012). In August 2012, Ohio, Indiana, and Kentucky signed the trading plan, making it the world's largest water quality trading program. At full scale, it could span up to eight states and potentially create a market for 46 power plants, thousands of wastewater utilities, and approximately 230,000 farmers. Water quality trading is a market-based approach that allows a discharge source to purchase reduction credits from another allowable source. In this case, tradable credits are generated by reductions in nutrient runoff achieved by implementing additional BMPs.

The pilot phase ran from 2012 to 2015 to provide an opportunity to test different trading mechanisms in advance of new or more stringent regulatory drivers.

8.7.3.1 Program components
EPRI initiated the trading program along with support from a multi-institutional collaboration and funded a feasibility assessment including in-depth reviews of 80 assessments and trading programs for guidance in developing the Ohio River Basin Trading Program. The assessment process, accounting methods, baseline requirements, and standards were developed within this network. It relies on a practice-based program based on the USDA conservation delivery system protocol. First, EPRI enters into agreements with the relevant state conservation and agriculture agencies to initiate the downstream flow of funding. Second, the state agencies enter into agreements with SWCDs that will periodically monitor, inspect, and verify the BMPs. Third, the SWCDs enter into agreements with the landowner to fund the implementation of BMPs. EPRI will own all of the credits that are established through these BMPs.

Nutrient reductions are estimated using an EPA model and the Watershed Analysis Risk Management Framework (WARMF) model (Neilson et al., 2003). Assurance, verification, and auditing of the trading credits included in the transaction process will be periodically inspected by an appropriate verifier. Verification records will be maintained and the nonconfidential portions of those records may be made available to the public upon request. Monetary value will be provided to farmers from a variety of potential commercial sources such as manufacturing, power plants, industrial processes, and wastewater treatment facilities that hold permits issued by local or regional regulatory agencies that allow discharges into water bodies (rivers, lakes, and streams).

8.7.3.2 Assessment and discussion
The EPRI relied on a mix of governance styles (H-4, M-3, and N-7) to develop and deliver the program (Table 8.9). EPRI used a network-lead governance style to organize, fund, and develop the program criteria. Its

delivery was based on a mix of hierarchy and market governance styles similar to the USDA conservation delivery system. Hierarchy governance was used in oversight. Valuation is based on water quality credits with payments sourced from commercial dischargers. Participants voluntarily choose to enroll in the program based on their perceived needs and benefits.

8.7.4 The Sustainability Consortium

TSC is an organization of diverse, global participants collaborating to build a foundation to account for consumer product sustainability and support methods for improvement. TSC engages corporate, academic, and governmental and nongovernmental organizations for their expertise and insights for developing this process. These include creating transparent methodologies, tools, and strategies to support a new generation of products and supply networks that address environmental, social, and economic objectives (TSC, 2015).

8.7.4.1 Project components

In July 2009, Walmart announced it was creating a "sustainability index" to address its product supply chains. TSC was formed in 2009 by a number of retail organizations and is jointly administered by Arizona State University (ASU) and University of Arkansas (U of A). It has an elected board of directors with corporate, civil society, and academic representatives and is managed by a global executive team.

The TSC network chooses the product categories to work on based on their estimated impacts and the volume of products sold. Stakeholders develop and use the Sustainability Measurement and Reporting System (SMRS) as a standardized framework for the communication of sustainability related information throughout the product value chain. Using this tool, companies can improve the quality of decision making about product sustainability and design better products. The reporting system is currently composed of a portfolio of three components: product category dossiers, category sustainability profiles, and key performance indicators (KPIs). These components provide a traceable, consistent, and science-based method for understanding the key environmental and social issues within the life cycle of a product category.

TSC does not verify KPIs but network participants choose to implement their own verification or certification process. TSC is developing a training and certification course to create a pool of service providers for the supply chains. Suppliers could hire the certified technician to provide assurance and assist in improving their product sustainability. The motivation for suppliers to improve on product sustainability is to gain retailer market share as their supply chains become more sustainable.

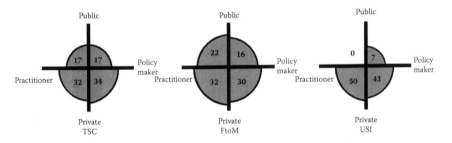

Figure 8.7 The governance actor participant percentage of TSC and FtoM studies are fairly rounded with participants from each sector. The USI case study was initiated by the private practitioner sector and was able to provide conservation technical assistance, rather than incorporate public practitioner actors.

8.7.4.2 Assessment and discussion

The TSC relies primarily on a network-administrative governance style (H-0, M-6, and N-8) to meet it objectives (Figure 8.7). A network-lead governance style was used to initiate the project with Walmart providing guidance. Funding for the consortium occurred in a market governance framework as stakeholders chose to participate and fund the effort based on the perceived benefits of joining. Network-administrative governance was used to develop the program with ASU and U of A, as external entities, administering and facilitating the effort.

Program delivery is based on market governance with stakeholders deciding to participate based on organizational benefits and interests. Oversight for the data sets and management is provided by the individual stakeholders using an open network-participant governance allowing organizations to determine the level of scrutiny. TSC is sponsoring a certification training program to provide a more uniform pool of certifiers with assurance credentials. The valuation type is an increase in market access for sustainable products. Both specific practices and outcome-based KPIs are used to account for sustainability outcomes of a product with improvement in product sustainability potentially leading to an increase in retail market access for that product line.

8.7.5 Field to Market

Field to Market is a collaborative stakeholder group of producers, agribusinesses, food and retail companies, conservation and nonprofit organizations, universities, and agency partners that are working together to define, measure, and develop a supply chain system for agricultural sustainability. A primary goal is to verify outcomes to enable supply chain sustainability claims relative to agriculture production. This involves creating metrics that represent sustained improvement of land use by increasing productivity, reducing greenhouse gases, and improving water quality, irrigation water use, and energy use (FtoM, 2016).

8.7.5.1   Project components

Field to Market was convened by The Keystone Center, a nonprofit organization specializing in collaborative decision-making processes for environment, energy, and health policy issues. A board of directors and five working groups established sustainability goals and developed assessment processes and metrics to measure against those goals.

The program delivery is through an online tool that agriculture producers input field data and management activities. The tool calculates index scores to determine their environmental impact. This data is also aggregated by region and commodity. Growers are responsible for the accuracy of their data. The organization promotes a voluntary, collaborative approach to sustainability that is expressly science-based, technology neutral, and focused on outcomes that are within a grower's control.

8.7.5.2   Assessment and discussion

The Field to Market governance ratio contains both market and network governance styles (H-0, M-7, and N-7) to achieve its objectives (Table 8.11). Under a network-lead governance style, the group was convened in 2007 and is facilitated by The Keystone Center, a neutral, nonprofit organization specializing in collaborative decision-making processes for environment, energy, and health policy issues.

It now functions using a network-administrative governance style where the Field to Market alliance manages the development of metrics to address agricultural supply chain sustainability for a broad spectrum of stakeholders. The delivery of the program relies primarily on a market governance style. Growers enter field management data to determine sustainability levels and address landscape conservation as they deem necessary. They are also responsible for the accuracy of their data. The value associated with participating is in being able to determine whether they meet the resource objectives as determined by the Field to Market criteria.

8.7.6   United Supplier's SUSTAIN

USI was established in 1963 as a cooperative by 30 Iowa retailers joining forces to manufacture feed for livestock. Today, USI is a member-owned wholesaler that provides agricultural products and services to about 700 grower cooperatives and retailers covering 45 million acres of farmland in the United States. In July, 2014 USI initiated the SUSTAIN® program to promote the critical concepts of right rate, timing, source, and placement of fertilizer (4Rs) to ensure higher nutrient use efficiencies and less nutrients and associated chemicals ending up in water and air.

The SUSTAIN program allows agriculture retailers to differentiate themselves in the marketplace by offering services to growers to improve efficiencies and sustain agricultural landscapes. Advisors, growers, and conservation staff will provide oversight to ensure practices meet the SUSTAIN goals.

8.7.6.1   Project components

USI created SUSTAIN in partnership with Environmental Defense Fund to develop and implement a fertilizer efficiency program for meeting sustainability supply chain objectives (Toot, 2014). Support and funding for SUSTAIN comes from, among others, Walmart, General Mills, Smithfield, and Coca-Cola Company.

The program is designed to track and verify implementation of 4R practices and create metrics for aggregated reporting for sustainable supply chains. USI also introduced a new private sector conservation planning initiative in May 2014 to deliver technical assistance related to applying conservation practices. The system adopted is similar to the traditional USDA conservation delivery system, but with a market-based approach. It is delivered and implemented through their network of retailer owners and staff.

The SUSTAIN program provides value to the growers through production efficiencies based on meeting voluntary resource objectives. The SUSTAIN program anticipates future values may be associated with corporate sustainability supply chains and carbon markets.

8.7.6.2   Assessment and discussion

The SUSTAIN program relies on mix of governance styles (H-5, M-4, and N-6) to achieve its objectives (Table 8.12). Using a network-lead governance style, the effort was initiated by USI with funding sources from several private sector policy makers. USI provided leadership in developing the SUSTAIN components with input from external stakeholders, particularly the Environmental Defense Fund.

The SUSTAIN 4R program is delivered through USI retail outlets through a hierarchy governance structure and data management process. Trainings are delivered by USI to retailers. The implementation of 4R practices is based on decisions by the growers as they deem appropriate. Oversight to ensure data meets the criteria of the corporate supply chains will be verified through a process developed by EDF and USI. Currently, meeting a voluntary resource objective is the only value associated with the SUSTAIN program, except for the potential efficiencies gained through productivity increases.

8.7.7 Group III discussion

The Group III case studies includes two (AWQCP and CBay) that are led predominantly by public governance actors and four (EPRI, TSC, FtoM, and USI) that are led predominantly by private governance actors as shown in Figures 8.8 and 8.9. Without further analysis, these governance compasses represent a significant shift from *government to governance* in agricultural sustainability in the last decade.

AWQCP and the CBay cases in Figure 8.8 rely on public governance actors, particularly public policy makers as the primary governance actor, and have adopted a hierarchy governance style and a practice-based accounting method. The EPRI study uses the unique mix of both private policy makers and public practitioners as the primary governance actors. It uses a network-lead governance style and a practice-based accounting system.

In Figure 8.9, the TSC, FtoM, and USI governance compasses show a reliance on the private sector as the primary governance actors and each adopted a network governance style. TSC and FtoM use network-administrative

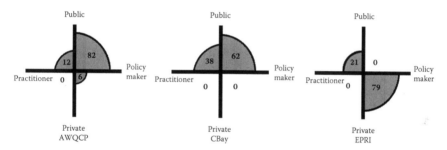

Figure 8.8 Primary governance actors of the AWQCP and the CBay are predominantly from the public policy-maker sector at 82% and 62%, respectively. The EPRI project was initiated by a nonprofit organization with stakeholders consisting primarily of government agencies and utilities.

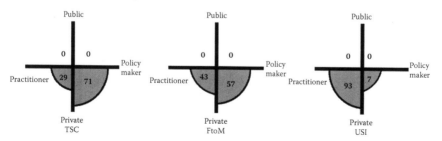

Figure 8.9 Primary governance actors of TSC, FtoM, and USI originated from private sector governance. TSC and FtoM were initiated by policy makers and the USI program by private practitioners.

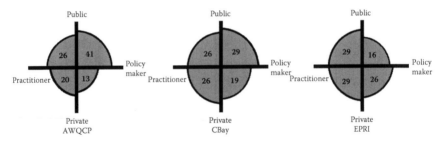

Figure 8.10 The governance actor participant percentage of the AWQCP, CBay, and EPRI studies are fairly rounded with participants from each sector.

and USI uses a network-lead governance style. TSC and USI rely on a combination of practice- and outcome-based accounting systems, with FtoM using an index-based outcome process.

Despite these varied sources of project leadership and primary governance actors, the total governance participants involved in each case were quite inclusive (Figures 8.7 and 8.10). Similar to the Group II case studies, in which a variety of primary governance actor strategies were used, the percentages related to the total governance participants is fairly well-rounded. The exception is the USI study that contracted with a private sector conservation business to address the resource assessment, planning, implementation, and assurance processes traditionally delivered by the local government agents (public practitioners associated with the USDA conservation delivery system).

8.8   Overall case study findings

Except for the now nonexistent SCS CDS, all the case studies had fairly inclusive governance participation by all the governance actor sectors, regardless of the sector providing primary governance. Nine of the eleven cases' primary governance actors reside in the policy-maker hemisphere. Five (SCS, NRCS, LEQA, AWQCP, and CBay) reside in the public policy-maker sector, and four (MMEQA, EPRI, TSC, and FtoM) in the private policy-maker sector. Two cases have primary governance actors in the practitioner hemisphere, with one (AgEQA) in the public practitioner sector, and one (USI) in the private practitioner sector. Due to the many potential combinations of preferred modes of governance, each governance framework was unique in some aspect.

Since governance styles are rooted in the culture of organization, the *choice* of which actors provide primary governance is often less about strategy and more about the culture of the organization leading the effort. A more pragmatic approach may be an assessment of the governance styles used in the components of the case studies. While choosing the type of governance style to be used for each component and subcomponent may not be a completely conscious decision, the decision may be more based on logistics and expertise of the participating entities. These decisions may be loosely, yet closely linked to which governance actors and which governance styles *end up* being chosen.

8.8.1 *Governance of project components*

Governance styles used for the components and subcomponents of the projects is the level at which meta-governors can guide efforts and influence outcomes. Table 8.13 lists each case study and identifies the primary governance style used for each component and the overall project governance style and accounting method used to plot the governance footprint.

8.8.1.1 *Development*

The development component was predominantly governed using network governance, particularly network-lead which relies on an entity internal to the effort to take charge. A network-lead governance style may also act

Table 8.13 Governance styles adopted for specific project components to create unique governance footprints

	Project components				Governance footprint	
	Develop	Deliver	Oversight	Value	Style	Account
Group I						
SCS	H	H	H	M	H	P
NRCS	H	H	H	M	H/m	P/O
Group II						
MMEQA	N-l	M	H	M	N-l	P
LEQA	N-a	M	H	M	N-a	P/O
AgEQA	N-p	M	H	M	N-p	O
Group III						
AWQCP	N-l	H	H	M	H	P/O
CBay	N-l	H	H	–	H	P
EPRI	N-l	M	H	M	N-l	P
USI	N-l	H	H	M	N-l	P/O
TSC	N-a	M	N-p	M	N-a	P/O
FtoM	N-a	M	M	M	N-a	P
Prominent	N-l	M	H	M		

Note: Commonalities, but not unanimous styles, emerged in the project components. Project development favors network-lead styles, delivery favors market, oversight hierarchy, and project valuation favors market governance. The governance footprint coordinates are listed in the right-hand columns.

as hierarchy governance, but incorporate consensus-based decisions. The SCS and NRCS used hierarchy governance to initiate and develop the conservation delivery system and five (MMEQA, AWQCP, CBay, EPRI, and USI) used network-lead governance styles.

Of the remaining three, TSC and FtoM used network-administrative and AgEQA used network-participant governance in the development phase of the program. These case studies suggest that strong hierarchy-type leadership with a particular vision is needed to initiate agriculture landscape sustainability projects.

8.8.1.2   Delivery

The delivery component had a mix of market (6) and hierarchy (5) governance with market governance a secondary influence in most of the hierarchy cases. The delivery components of collecting data and information, data management, planning, implementation, and training can be extensive and expensive. Since market governance supports an open process, a potentially unlimited number of actors are eligible to contribute and innovate to find and locate the most cost-effective and efficient means to accomplish the tasks.

8.8.1.3   Oversight

The oversight component used hierarchy governance with the two exceptions of TSC and FtoM using network-participant and market governance, respectively. It should be noted that TSC is in the planning phases of a training and certification program to provide expertise for assessing supply chain sustainability. Adding new actors with new roles has the tendency to shift governance styles and affect governance frameworks if the process becomes institutionalized. As these new strategies are implemented, new processes surrounding data acquisition and assurance may emerge.

8.8.1.4   Valuation

The valuation component was unanimously market governance except for the CBay project that did not place a value on the conservation outcomes and data they sought. In the 10 cases offering values, individuals decided if the incentives met their needs and if so, pursued transactions to exchange data and value.

In general, the majority of the agriculture sustainability efforts are initiated with a top-down approach and developed with the input from a broad group of stakeholders using network-lead governance. The delivery component of the cases is addressed with a mix of governance styles, or a meta-governance strategy with a prominent market governance presence. Project oversight is conducted with a top-down, organized process of checks and balances using hierarchy governance. The valuation component relied on a market governance style.

8.8.2   Governance footprints

A *governance footprint* is a representation of the ratio of the governance styles in the overall framework (Meuleman, 2008). In these cases, the governance footprints were plotted using the governance style(s) relative to the accounting methods employed (Figure 8.11). The *y*-axis represents the practice-based, outcome-based, and hybrid or combination accounting methods. The *x*-axis represents the hierarchy and market governance styles and the three network governance types: lead, administrative, and participant.

The Group I case studies (ovals) illustrate a shift from *government to governance* to address the increasing complexity of agriculture sustainability. This relatively slight shift from hierarchy governance toward market governance occurred as the NRCS incorporated the TechReg and private sector conservationists. The NRCS also adopted outcome-based accounting methods for the eligibility and incentive components of (CRP, CSP) conservation programs. The SCS CDS governance footprint is plotted on Figure 8.11 as a practice-based accounting system and a hierarchical governance style. The NRCS CDS governance footprint is predominantly practice-based with a hierarchical governance style with a slight shift toward a combination/ hybrid-based accounting system and market governance.

The three Group II case studies (trapezoids) illustrate an evolution of pilot projects and a shift in network governance styles and accounting strategies. The earlier 2001–2009 MMEQA adopted a network-lead

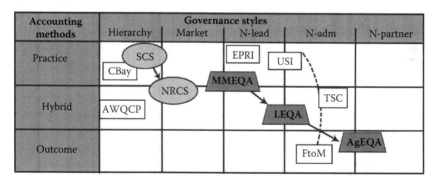

Figure 8.11 Governance footprints of the case studies are represented individually and by groupings. Group I (ovals) consists of the SCS and NRCS CDS with a shift in governance and accounting methods toward market governance and outcome-based accounting. Group II (trapezoids) consists of three state projects, MMEQA, LEQA, and AgEQA, that shift toward more inclusive governance styles and outcome-based systems. Group III (rectangles) are independent case studies with a variety of governance strategies and accounting methods. The dotted line connecting USI, TSC, and FtoM represents formal and informal strategies to integrate efforts.

governance style with a practice-based accounting system. The 2010–2011 LEQA adopted a network-administrative style with a hybrid accounting system using practice- and outcome-based systems. The 2013–2014 AgEQA shifted toward a network-participant governance style and fully adopted an index-based accounting system. The MMEQA governance footprint is plotted on Figure 8.11 as a practice-based accounting system and a network-lead governance style. The LEQA governance footprint is plotted on Figure 8.11 as a combination/hybrid-based accounting method and a network-administrative governance style. The AgEQA governance footprint is plotted on Figure 8.11 as an outcome-based accounting method and a network-participant governance style case.

The six Group III case studies (rectangles) seem generally scattered with a mixture of governance styles and accounting methods. One could create two subgroups with CBay, AWQCP, and EPRI in the first subgroup. CBay uses hierarchy governance with practice-based accounting and resembles the SCS footprint. AWQCP uses hierarchy governance with a hybrid accounting strategy using both practice- and index-based criteria. EPRI uses network-lead governance and practice-based accounting (Figure 8.11).

The three remaining case studies, USI, TSC, and FtoM, are shown connected with a dashed line to represent collaborations between TSC and FtoM (Field to Market, 2014) and less formal collaborations linking USI with corporate and NGO efforts. TSC relies on network-administrative governance and uses key performance indicators based on practices and outcomes. FtoM uses network-administrative governance and indices as an outcome-based accounting system. USI's SUSTAIN program is using a network-lead and a practice-based model with the use of metrics to aggregate data at a larger scales (Figure 8.11).

8.9   Aligning governance actors and styles

A primary purpose of analyzing these case studies is to understand how organizations and their partners approached agricultural landscape sustainability and which actions and strategies aided in aligning the activities to achieve sustainability objectives. A few observations include the following:

- There is a tendency for the projects to be initiated by a network-lead, policy organization that is able to collaborate with several stakeholders to create consensus.
- A common strategy is to offer a value to those practitioners that are able to achieve the outcomes desired by the policy organizations.
- The delivery mechanisms of the projects are generally incentive-based and use market governance.

- Oversight is usually provided by the lead organization using hierarchy governance.
- A mix of accounting strategies is used with a trend to move toward or include a hybrid or outcome-based process.
- Overall, governance tended to shift away from hierarchy governance to the more innovative and inclusive market and network governance styles.

This trend toward network governance styles and outcome-based accounting methods provides governance actors with more opportunities to interact and align their activities to achieve interdependent outcomes. This shift toward network governance allows more interconnections to occur than was previously possible with hierarchical governance (Fisher, 2006). Swihart (2009) suggests complex organizational efforts must be centered on the *point of service* so the outcomes are clearly defined by those who produce the outcomes.

This presents the question of whether it is more advantageous to seek convergence of governance styles and accounting strategies, or design a collaborative process to transcend the many governance footprints of the case studies and others too numerous to mention. Meuleman (2013) states a transdisciplinary approach with open innovation and greater compatibility is needed to achieve sustainability objectives. Scharmer and Kaufer (2013) used the term "co-creative eco-system" to describe a platform and a space for cross-sector innovation and engagement. It is from this perspective that a shared governance platform is proposed.

chapter nine

A shared governance platform

Governance styles of organizations do shift and perhaps even change over time, but organizations should not be expected to adopt different governance styles for the sole purpose of alignment. Governance is a cultural-based characteristic, and so, highly resistant to change (Meuleman, 2008). And in any case, no one governance style is optimum for achieving all objectives as each has their strengths and weaknesses for achieving particular objectives (Jessop, 2003). To resolve this governance paradox, a *shared governance* platform is proposed to enable governance actors from organizations with different governance styles to interact to achieve common outcomes.

9.1   Shared governance

Shared governance is not defined as a new bureaucratic system or government program but a collaborative process to engage traditional, sector-based organizations often constrained by their geographical boundaries and singular purpose (Marsh, 2000). Porter-O'Grady (1992) states shared governance causes a shift from a command-and-control (governance) to a knowledge-based decision system. He states shared governance creates a platform for open engagement with all (actors) contributing to the outcome and having ownership of their decisions. Shared governance is not a democracy, it is an accountability-based approach to structure in which there is a clear expectation that all members of a system participate in its work.

Scharmer and Haufer (2013) used the term "cocreative ecosystem" to describe an evolution of organization structure which is similar to the shared governance concept. He defined it as a platform and a space for cross-sector innovation and engagement. In the evolutionary process of governance styles, shared governance is a relatively recent occurrence (Figure 9.1) and has emerged as social issues have become more complex.

9.1.1   Shared governance principles

In simple terms, shared governance is shared decision making based on the principles of partnership, equity, accountability, and ownership at the point of service (Porter-O'Grady, 2001). These four general principles of

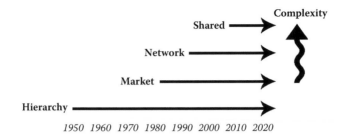

Figure 9.1 Shared governance is added to the evolutionary governance timeline of Figure 5.1. Versions of shared governance were used in the university system as early as the 1960s (Olson, 2009) and emerged in the healthcare field in the 1990s (Porter-O'Grady, 1992). (Adapted from Meuleman, L., *Public Management and the Metagovernance of Hierarchies, Networks and Markets: The Feasibility of Designing and Managing Governance Style Combinations*, Heidelberg, Physica-Verlag, 2008, 43).

shared governance empower actors with decentralized decision-making ability at the point of service (Swihart, 2006).

1. *Partnerships* are essential to building relationships and involve all stakeholders in decisions and processes. It implies each member has a key role in achieving landscape sustainability.
2. *Equity* implies each stakeholder is as important as the others, but does not imply that each stakeholder is equal in terms of scope of practice, knowledge, authority, or responsibility.
3. *Accountability* is the basis for responsibility and allows for evaluation of performance. It creates a willingness to invest in decision making and express ownership in those decisions.
4. *Ownership* requires all stakeholders to contribute something, to own what they contribute, and to participate in achieving the purpose of the work.

These shared governance principles are compatible with multisided platforms: an emerging business model that creates new value through stakeholder interaction.

9.2   *Multisided platforms*

Multisided platforms create value by bringing two or more customer types together and facilitating interactions to solve a transaction cost problem that otherwise makes it difficult or impossible for different groups to get together (Evans and Schmalensee, 2012). These platforms play critical roles in industries such as credit card payments, mobile phones, financial exchanges, advertising, and various Internet-based industries (Hagiu and Wright, 2011).

Since the emergence of multisided platform businesses in the last decade, many definitions have been proposed to explain their network effects and characteristics. Hagiu and Wright (2011) define a multisided platform to be an organization that creates value *primarily* by *enabling direct* interactions between two (or more) distinct *types* of *affiliated* customers. Others define these platform markets as two groups of agents who interact where one group's benefit from joining a platform depends on the size of the other group that joins the platform. Platforms can also serve as foundations upon which other firms can build complementary products, services, or technologies (Armstrong, 2006).

Evans and Schmalensee (2007) referred to multisided platforms as an *economic catalyst*: creating value that could not exist or would be greatly reduced without it. This new value is created as a result of solving coordination and transaction cost problems between the groups (Evans and Schmalensee, 2012). Evans (2011) stated platforms usually perform three core functions to some degree: matchmaking to facilitate exchange by making it easier for members of each group to find each other, building audiences, and providing shared resources to reduce the cost of providing services to both groups of customers.

Choudary et al. (2015) state a multisided platform is a plug-and-play business model that allows multiple participants (producers and consumers) to connect to it, interact with each other, and create and exchange value.

The fundamental role of a multisided platform is to enable parties to realize gains from trade or other interactions by reducing the transaction costs of finding each other and interacting. Multisided platforms coordinate the demand of distinct groups of customers who need each other in some way. From this perspective, multisided platforms provide a virtual or physical meeting place for customers and users to create value (Evans and Schmalensee, 2012) by reworking the economics of participation (Evans, 2008). This reworking creates efficiencies by reducing *search* and *transaction costs* by bringing consumers and producers together (Hagiu and Wright, 2007).

Multisided platforms emerge in situations in which transaction costs, broadly considered, prevent two sides from working directly with each other (Evans, 2011). Platforms overcome this obstacle by creating a *network community* where users can interact within a *technological infrastructure* and allow users to interact with a value-enhancing *database* (Choudary et al., 2015).

9.2.1   Network community

The network is the individuals and groups associated with the use of a multisided platform. In multisided platforms, this association creates an *indirect network effect*: an increase in the value that a customer on one side realizes as the number of customers on the other side increases (Evans, 2011).

These indirect network or cross-side network effects allow some multisided platforms to set the price of participation below cost on one side of the platform to attract customers that in turn attract customers and users on another side of the platform to whom they charge prices above cost. For example, a credit card may be offered free to consumers to entice them to sign up and then businesses may be charged the full cost of making the transaction (Hagiu and Wright, 2011).

Because the businesses' benefits increase as the number of cardholders increase, they are compelled to accept the credit card and pay the expenses.

Network communities are more prone to grow as friction is minimized for users to connect to the network. Social networks such as Facebook and LinkedIn gained initial traction through incorporating network member contact lists to reduce sign-up friction (Choudary et al., 2015). Full network effects are achieved only after a certain critical mass is reached.

The increase in the use of platforms is aligned with the emergence of three transformative technologies associated with network connections: social networks, cloud computing, and mobile access. Social networks connect people globally and maintain their identity online. Cloud computing allows anyone to create content and applications for a global audience. Mobile access allows connection to this global infrastructure anytime, anywhere. The result is a globally accessible network of entrepreneurs, workers, and consumers who are available to create businesses, contribute content, and purchase goods and services. To create value in a network, a technological infrastructure is needed (Choudary et al., 2015).

9.2.2   Technological infrastructure

The technological infrastructure of a multisided platform is the means to get a job done, to solve a problem, or to provide a fix. The technological infrastructure delivers the fundamental services the platform needs to perform for those customers who are critical to the success of the platform (Hagiu, 2007). It is the built component of the platform. It is also the component that external producers build on. Examples of infrastructure and producers include Android® and the developers building apps, YouTube® and the producers uploading videos, and eBay® and the sellers posting items for sale (Choudary, 2013).

Baldwin and Woodard (2009) find the fundamental *architecture* behind all platforms is essentially the same. The system has a set of core, long-lived stable components with a set of periphery components that change over time. With this structure, economies of scale can be realized at the core due to fixed costs and increasing the production of the stable components. Economies of scope can also be realized by the periphery components as producers can experiment and choose the best outcomes but without compromising the core components.

Choudary et al. (2015) use the acronym TRIE (tools, rules, interaction, and experience) to describe the makeup and use of the technological infrastructure. *Tools* include the technology and the interface components as part of the platform, such as video uploading capabilities. *Rules* create the boundaries of user behavior. For example, limiting Twitter users to 140 characters is an obvious rule of use where a less obvious example is the use of algorithms determining search results. Tools and rules are what is built into and controlled by platform builders and it creates the infrastructure on which interactions occur and the conditions for such experiences to take place.

9.2.3 Database and content

Data is the foundation of any platform business. Data is increasingly becoming the new currency for platforms (Shields, 2015). Decisions on how much data to store, who to share it with, who not to partner with, what data to make available, and what data to monetize are essential for a multisided platform. In most cases, data serves to provide relevance: matching the most relevant content, goods, and services with the right users. In other cases, the value may lie in the data layer as a data-intensive platform, where the value is entirely in the data being aggregated. Most multisided platforms go beyond simple aggregation of databases and *curate* data and those providing it.

Curation is to ensure the quality of the data is adequate and it is typically done in one of three forms: editorial, algorithmically, and social. Using editors to curate data is effective only to a certain scale unless the editorial function is moved out into the network community by educating the community on the tools of curation (rating, review, reporting, etc.). A second method is using algorithms provided by editors or the community and hence scaling algorithmic curation works very closely with scaling the size of the community. Social curation is relying on the community to curate data and may rely on participant reputations or opinions of experts over novices.

9.3 Multisided shared governance platform

Merging the concept of a shared governance platform with that of a multisided platform creates a multisided shared governance (MSSG) platform. It is designed to operate as a multisided business platform *and* to address governance issues related to agriculture landscape sustainability. The MSSG is a geographic-based platform that acts as the catalyst for the business, social and governmental aspects of agricultural landscape sustainability. It allows government, nongovernmental organizations (NGOs), and the private sector to interact and achieve their interdependent objectives. A MSSG platform provides a new *space* for organizations with conflicting governance styles, disparate valuation strategies, and accounting methods to interact.

In this new space, the network community is able to interact using a geographic information system (GIS)-based technological architecture that accounts for natural capital outputs using landscape index data. This creates a new supply that entices new user behavior resulting in new transactions. New supply, behaviors, and transactions rework the economics of participation in agricultural landscape sustainability.

9.3.1   Creating a new supply

The MSSG platform creates a new supply by delineating the landscape into equal-sized units called natural capital cells (NCCs) as described in Chapter 7. Each NCC has a unique identification code representing its latitudinal and longitudinal location and contains the landscape data required to calculate landscape management index values. Platform users access an *index depository* to choose the index representing the ecoservice of interest. These calculations produce a natural capital unit (NCU), a potential asset of the NCC.

An NCC containing ID number, site, and management data and three NCU values generated by a soil conditioning index (SCI), water quality index (WQI), and a habitat suitability index (HSI) is shown in Figure 9.2. The indices included were developed by the U.S. Department of Agriculture (USDA) and are valid examples, but the indices used may also be developed by platform users. Indices included in the depository will undergo a curation process to ensure they are scientific-based, follow platform protocols, and have the capacity to adapt to the evolution of the platform.

NCUs may represent natural capital outputs and outcomes such as water purification, pollination potential, and carbon sequestration. In reality, these *supplies* are not new, as these outputs and outcomes have always existed, but economically speaking, NCUs are a new supply of tradable ecoservice units. NCUs become a tradable unit by a *conversion of sorts* from nontradable public and common goods, such as sequestered carbon or purified water, into NCUs that become tradable private goods and club goods. This *conversion* is critical, as public and common goods are not tradable because as goods they are nonexcludable.

Excludability lends itself to ownership, and without ownership, assets (such as ecoservices) become *dead capital* unable to generate returns over and above that associated with their direct use (Landell-Mills and Porras, 2002).

Figure 9.3 is a textbook version of the different categories of goods and services relative to their excludability and rival status. Bollier and Helfrich (2012) state the classification of a certain good is not inherent to the good, but is created through a variety of social norms and institutional

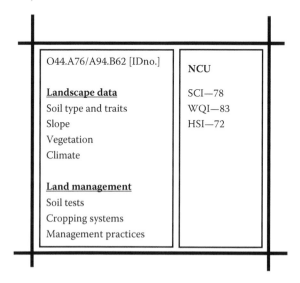

Figure 9.2 A natural capital cell (NCC) contains a location identifying number, landscape, and management data and access to an index depository to calculate natural capital units (NCUs). The NCUs generated by the NCC resource characteristics and management activities create a new supply of tradable ecoservice units.

	Excludable	Nonexcludable
Rival	Private goods	Common goods
Nonrival	Club goods	Public goods

Figure 9.3 Economic goods are categorized relative to their excludable and rival status. Goods deemed nonexcludable do not have ownership characteristics and markets do not spontaneously develop.

approaches that create classifications that then *seem* natural. Ecoservices, relative to NCU data, could exist in more than one classification in Figure 9.3 depending on the capacity of the MSSG platform to manage the data in excludable and rival manners.

9.3.1.1   Nonexcludable goods

A nonexcludable good is a good whereby it is not possible to exclude or prevent people from using the good. Nonexcludability means that consumers cannot be prevented from enjoying the good or service in question, even if they do not pay for the privilege. Today, for instance, it is currently difficult, if not impossible, to exclude downstream communities from benefiting from improved water quality associated with forest regeneration upstream.

Nonexcludable goods are categorized as *public goods* and *common pool goods* with their differences being nonrival and rival, respectively.

9.3.1.1.1   Public goods   Public goods have two salient properties, namely, nonexcludable in supply and nonrival in consumption. A lighthouse signal is a classic example of a public good, where the provision is both nonrival and nonexcludable. Public goods contrast with private goods that are excludable and rivalrous in consumption and can be sold to those who can afford to pay the market price (McNutt, 1999).

Where goods are nonrival, the consumption of a good or service by one individual does not reduce the amount available to others. In this situation, there is no competition in consumption since an infinite number of consumers can use the given quantity supplied. An example of a nonrival ecoservice is carbon sequestration. Once carbon is sequestered, the global community may benefit from this in terms of a reduced threat of global warming. Every individual can enjoy the benefits regardless of if they pay for the benefit. Where nonexcludability and nonrivalry exist, they undermine the formation of markets since beneficiaries of the good or services have no incentive to pay suppliers. As long as an individual cannot be excluded from using a good, they have little reason to pay for access. The result is free riding: the ability to use a resource based on others' activities and payments (Kim, 1984). Where everyone adopts free-riding strategies, willingness to pay for public goods will be at or near zero, and the product or service may not be supplied.

9.3.1.1.2   Common-pool goods   Goods can be characterized by varying degrees of nonrivalry and nonexcludability. The extent of nonrivalry and nonexcludability will determine the degree of market failure. For instance, where goods are nonexcludable, but rival, they are described as common-pool or open-access resources. They are rivals because there are limits to the available supply or limits to the capacity to supply.

Garrett Hardin's 1968 essay, "Tragedy of the Commons" describes an open-access or common-pool system, where unlimited access to the limited resource of the pasture results in a degradation of the resource. More modern examples of agricultural common-pool resources include irrigation or drainage networks used by local farmers. Elinor Ostrom's (1999) research on open-access governance showed, in some cases, local cooperative actions were able to avert the "tragedy of the commons," but nonexcludable, rival goods are prone to over-consumption and degradation.

9.3.1.2 Excludable goods

A good is excludable if it is possible to prevent consumers who have not paid for it from having access to it. A good becomes excludable, not solely by its nature, but when one can at low cost prevent those who have not paid for the good from consuming it. For example, it is cost-effective to require people to pay for a stamp before the postal service will deliver mail or to require a ticket before they board a train. In contrast, it is generally not cost-effective or easy to prevent people from entering a park or from listening to a radio station.

Likewise, it is currently difficult or not cost-effective to exclude people from using or enjoying ecoservices generated from the agriculture landscape. For ecoservices to become a tradable good, they must be excludable, or essentially *converted* to an excludable good. This property of excludability in the supply of a public good is the *sine qua non*, an absolute condition for club goods (McNutt, 1999).

9.3.1.2.1 Club goods The salient characteristic of a club good is excludability (McNutt, 1999). Goods that are excludable and nonrival are described as toll or club goods since markets can be set up in the form of tolls. An example of a toll good is that of roads in national parks, where entry is controlled. Each consumer, theoretically, is able to enjoy the good without reducing its overall value for others, until congestion occurs. Examples of club goods include cinemas, cable television, access to copyrighted works, and the services provided by social or religious clubs to their members. Public goods with benefits restricted to a specific group may be considered club goods. Therefore, a public good that becomes excludable is a club good (McNutt, 1996). In many respects, this club provision offers an alternative to central government delivery and management of public goods. Clubs organize to enable members to exploit economies of scale in the provision of the public good and to share in the cost of its provision (McNutt, 1999). Excludable club goods become private goods when their nonrival status becomes rival.

9.3.1.2.2 Private goods Once property rights are established, the good eventually becomes an excludable and rival private good. Where goods are both excludable and rival, they are described as private as they may be easily supplied by the private sector based on market transactions. A bushel of grain is easily measured, stored, transported, and ownership transferred. One can easily exclude others from using the bushel of grain, as it is rival: if one consumes the bushel of grain, no one else is able to use it.

9.3.1.3   *"Converting" nonexcludable goods to excludable goods*

The logic or strategy of creating a new supply of tradable ecoservices resides in the *conversion* of nonexcludable ecoservices into a private good and/or a club good. This conversion is accomplished, not by addressing the excludability of the use of the actual good, but by creating an excludable good based on the data that represents the production of the ecoservices. In other words, the NCU, not the water purifying ecoservice or the carbon sequestration ecoservice, becomes the tradable good. Since the economic classification of a certain good is not inherent to the good, but created through a variety of social norms and institutional approaches, changing the norms and institutional approaches can *convert* a good from one economic classification to another (Bollier and Helfrich, 2012).

This conversion is enabled due to ownership of the data required to calculate and substantiate NCU values. NCUs are derived from calculations composed of landscape data and land management information. Since land management information can be proprietary, a NCU calculation, in part, could also be owned by the land manager. This excludable NCU has the potential to be valued as either a private good (rival) or a club good (nonrival). As a club good, it could be traded with several entities seeking landscape sustainability data for the purpose of accounting for sustainability of a supply chain or watershed.

As a private good, the NCUs could be sold to a specific entity for ecoservice credits or for natural resource mitigation. Entities making sustainability claims would presumably need the data to substantiate those claims. As an exclusive good it restricts free-ridership and creates new user behaviors on the demand side.

9.3.2   *New user behaviors*

The new ecoservice values embedded in the proprietary NCUs enable sustainability demanders to value agricultural landscape sustainability in a manner not previously available. These new values instigate new behaviors from those demanding sustainability. New behaviors from a corporation may consist of valuing the production of NCUs to address a particular resource issue related to the commodities they purchase. A retailer can exert their influence on the type of NCUs needed to access their market. Government agencies may identify the NCUs needed to enroll in a program or to meet a water quality standard. Utilities may identify the types and quantities of NCUs needed to be eligible to engage in water quality trading.

In all cases, the demander readily identifies the NCU supply it desires along with a criteria or value associated with the NCU. The NCU and the economic value associated with it becomes a *sustainability market signal*. For example, a retailer could state that it will purchase a food product that has a WQI > 80. Those in the upstream market *hearing* this market signal will inquire about a WQI and all the issues surrounding the product to determine how it may meet and maintain this market criterion. These new supplies and values instigate new behaviors in pursuit of new transactions. It is this market signal concept being developed and applied by The Sustainability Consortium members and others to encourage suppliers to account for and improve their supply chain.

9.3.3   Enabling new transactions

A primary objective of the MSSG platform and multisided platforms, in general, is to enable transactions that normally would not occur. For example, OpenTable is an application platform connecting potential customers with restaurants with readily available seating (Hagiu, 2014). Without the platform, customers would have to call, restaurant owners would have to answer the call, and then they would discuss the seating and waiting situation and if the match does not occur, the process would start over. Obviously, this is not an insurmountable problem, but the hassle to make and answer the calls becomes significant enough to encourage potential diners to seek out a more streamlined process using OpenTable. In the case of ecoservice markets, the transaction hurdles are not just an inconvenience, but are insurmountable as no ecoservice market has yet to mature in a typical market fashion.

9.3.3.1   Identify transaction costs

The crux of any economic transaction is the necessity for the transaction costs to be sufficiently less than the value in the exchange. A comparative advantage is created in a trade if the transaction costs are minimized in one trade relative to another trade of identical value. Ecoservice markets often have higher transaction costs due to their unique and challenging assessment and assurance needs.

Transaction costs include ex-ante (upfront) costs associated with obtaining relevant information needed to plan, negotiating agreements, making side-payments to gain agreement, and communicating. Ex-post costs are associated with monitoring performance, sanctioning and governance, and renegotiation when the original contract is unsatisfactory. Transaction costs are not only financial. Time and other in-kind contributions should be measured and, wherever possible, monetary values of these inputs should be calculated (Abdalla, 2008).

In addition to typical transaction costs, setup costs for water trading markets are typically high and span several years. Unavoidable costs include concept review and approval, baseline assessments, setting objectives, developing the market, creating the pricing structure, and securing stakeholder buy-in. Trading in any market will not occur if the transaction costs exceed the benefits of a potential trade (Brown et al., 2007).

9.3.3.2 Reduce transaction costs

Strategies that reduce transaction costs generally cause an increase in trades and provide opportunities for new markets to emerge. In "The Nature of the Firm," Coase et al. (1993) explain that firms exist because they reduce the transaction costs that emerge during production and exchange, and capture efficiencies that external entities or individuals cannot.

The intention of creating an MSSG platform is, in part, related to its capacity to reduce transaction costs. Transaction costs are reduced by creating a tradable unit (NCU) where one did not previously exist, by providing an integrated GIS-based accounting system, and like other platforms, transaction costs are reduced due to its matchmaking capacity. After reducing *total* transaction costs, addition savings per individual stakeholders may occur by *sharing* or dividing the transaction costs among stakeholders that mutually benefit.

9.3.3.3 Shared transaction costs and values

If transaction costs, in total, are not further reducible, it is still possible to reduce the transaction costs by sharing or dividing the transaction costs among multiple parties in interdependent transactions. This is possible due to the multiplicity of natural capital outputs from a single parcel of land. For example, if a unit of land (NCC) is managed to sequester carbon, purify water, and provide pollination habitat, the MSSG platform makes it feasible to account for each of these values (NCUs) singularly or in combinations at various scales.

It is this system complexity that *causes* high transaction costs, but if managed, the system's complexity can contribute to this unique strategy to reduce transaction costs. These transaction costs become sharable because ecosystems produce multiple outputs that are used by multiple stakeholders. In this scenario, NCU values become sharable. As long as multiple entities such as government, corporation, utilities, and NGOs mutually benefit from the output and outcomes of a single parcel of land, it is reasonable, and perhaps an economic necessity, to apply a transaction process that captures all the mutually beneficial values.

The shared transaction process supports the *shared value* strategy promoted by Porter and Kramer (2011). The Harvard economists describe shared value as a process, where corporate (and other) stakeholders generate and incorporate new values beyond financial goals. These shared values are embedded in transactions referred to a *Sustainability 1.0* and *1.5* and *Sustainability 2.0* transactions.

9.4 MSSG platform transactions

Ecoservice transactions are enabled by a gridded, GIS-based platform with the capacity of user groups to delineate areas of interest, generate NCUs, and interact with other user groups, particularly land managers, to acquire additional land management data, and assurance for specific outputs and outcomes. The MSSG platform supports three types of transactions: Sustainability 1.0 (S1.0), Sustainability 1.5 (S1.5), and Sustainability 2.0 (S2.0) transactions.

9.4.1 Sustainability 1.0

S1.0 transactions occur between a sustainability demander, such as a business in a sustainable supply chain, and the platform managers. S1.0 transactions are based on NCUs generated from superficial landscape data that is often readily available through government databases, satellite imagery, and GIS systems. The NCU indices and data sources for S1.0 transactions are shown in Figure 9.4.

S1.0 transactions are based on a NCU as an excludable and nonrival club good: excludable in the manner that the NCU is privately owned (by the platform managers) and nonrival in that it can be procured by multiple stakeholders to address each of their sustainability objectives. It is presumed that S1.0 transactions would involve data exchanges for relatively low costs.

9.4.2 Sustainability 1.5

S1.5 transactions occur between a sustainability demander, such as a business in a sustainable supply chain, and land managers. S1.5 transactions are based on NCUs generated from superficial landscape data (S1.0 data) and land management data. Figure 9.4 identifies the source of the NCUs for S1.5 transactions. NCUs generated for S1.5 transactions contain specific data that can provide more precise NCUs values.

S1.5 transactions are also based on an NCU as an excludable and nonrival club good: excludable in the manner that the S1.5 NCU is privately owned (by the land manager) and nonrival in that it can be procured

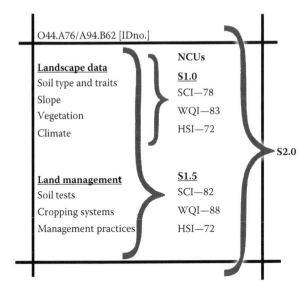

Figure 9.4 NCUs are generated within specific identified NCCs and transacted under three types. S1.0 trades are based on superficial landscape data embedded in the multisided shared governance MSSG platform. S1.5 is based on S1.0 data and proprietary land management information inputted by a platform user (land manager). S1.0 and S1.5 may be traded as club goods. S2.0 trades are based on quantified tradable credits based on S1.0 and S1.5 data along with more specific contract obligations and traded as private goods.

by multiple stakeholders to address their sustainability objectives. It is presumed that S1.5 transactions would involve data exchanges with slightly higher costs relative to S1.0 transactions.

9.4.3   Sustainability 2.0

Sustainability 2.0 transactions are based on NCUs generated from land management strategies designed to generate natural resource credits. As credits, S2.0 transactions are an exchange of an excludable and rival private good: excludable in the manner that the data is private and rival in that only one stakeholder can secure ownership. In this case, the demander is interested in the production of a specific ecoservice credit such as a quantified unit of water quality, wildlife, or carbon, with sole ownership. The MSSG platform would identify the specific area, practices applied, period of time, and other contractual obligations for the transaction. S2.0 credits are based on quantified goods. This is in contrast to S1.0 and S1.5 transactions that are based on the sustainability level of the production system by accounting for ecoservice values.

9.5   Symbiotic demand versus conflicting governance

An objective of creating a shared governance platform was to address the wicked component of the inherent conflicts of organizational governance styles. The MSSG platform transcends organizational governance conflicts by supporting shared governance principles at the point of service: the virtual and real landscapes. In this new space, governance actors are able to interact and symbiotically achieve their objectives. The MSSG platform is the catalyst that allows disparate stakeholders to interdependently align their activities to convert governance conflicts into *symbiotic demand*.

9.5.1   Disparate stakeholder governance

Even with common objectives, conflicts arise from the modes of action and the sets of values held by typical governance styles. Hierarchy governance values are based on the expectation that there should be a subordinate to the hierarchy, and this governance style relies on regulations and control instruments to meet goals. Market-based governance values a customer perspective and relies on competition and innovation to achieve results. Network-based governance seeks partners and cocreators and relies on trust to achieve outcomes (Meuleman, 2008). In this scenario, a multitude of governance strategies converge creating not just conflicts among the demanders, but inefficiencies and frustration for the supplier: the land manager.

Figure 9.5 includes six of the case studies applied to a fictional, yet realistic and probable scenario. It is possible and quite feasible that among the dozens of agricultural sustainability projects, a land manager could be expected to participate in a federal program (NRCS), a state water quality program (AWQCP), a regional watershed (CBay), a local agriculture retailer (USI), a corporate sustainability supply chain (TSC), and a water quality trading program (EPRI).

The NRCS and AWQCP cases in the top row use hierarchy governance and the remaining projects use network-lead governance styles. None of the projects use identical accounting systems. Two of the accounting systems are practice-based (CBay, EPRI), and the remainder use a combination of outcome- and practice-based methods. All of the cases have interest in S1.0-type values and are seeking S1.5-type values. The EPRI project also seeks S2.0-type (credit) values.

It is within this context that multiple stakeholders seeking common and complementary objectives can cause conflicts and inefficiencies. These conflicts are not necessarily based on *what* should be done, but on *how* things should be done. The result is low and diffuse ecoservice values accompanied by high and multiple transaction costs.

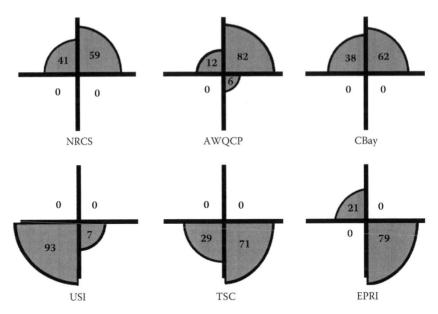

Figure 9.5 Six case studies illustrate the potential for conflicts and inefficiencies created when multiple organizations seek common sustainability objectives with programs developed from different governance sectors with differing governance styles. The three programs at the top of the figure were developed predominantly from public sector governor actors and the bottom three programs were developed predominantly by private sector governance actors. The Natural Resources Conservation Service (NRCS) and Agriculture Water Quality Certainty Program (AWQCP) use hierarchy governance originating from the public policy-maker sector. The remainder use network-lead governance with the CBay originating from the public sectors, the USI from the private practitioner sector, and TSC and EPRI originating from the private policy-maker sector. All use different accounting systems. Each seeks S1.0- and S1.5-type values and the Electric Power Research Institute, Inc. (EPRI) seeks S2.0 values.

9.5.2   Symbiotic demand

Symbiotic demand can emerge from the multiple-demand scenario described in Figure 9.5 *if* the governance actors from the various organizations can interact on a shared governance platform. Symbiotic demand is the result of multiple entities demanding sustainability values within a common forum so that the low and diffuse ecoservice values can be readily combined and the high transaction costs can be reduced and shared.

Symbiotic demand is different from the effect of economies of scale, where the costs per unit are decreased due to efficiencies of production and distribution. Symbiotic demand occurs when the cost *and the unit* is shared among users. The users can share a S1.0 and S1.5 NCU, as it is traded as a club good. Unlike a typical commodity where an increase in demand causes the price to increase and the single highest bidder receives the commodity unit, the cost of S1.0 and S1.5 values (per demander) may decrease as the number of demanders increases, and each bidder may receive the S1.0 and S1.5 values.

The MSSG platform enables multiple entities to procure the same S1.0 and S1.5 data (NCUs) to substantiate their sustainability claims or to meet their objectives interdependently. Figure 9.6 illustrates the flow path of NCC data to generate NCU calculations that are transferred to demanders. In this case, the MSSG platform generates S1.0 WQI data based on NCC landscape data. To generate S1.5 WQI data, land management data from the NCC is added to S1.0 WQI.

The calculations can occur somewhat spontaneously as the NCC interfaces with the indices chosen from the platform index depository. The calculation and assurance of S2.0 WQ credit is developed through a contractual basis.

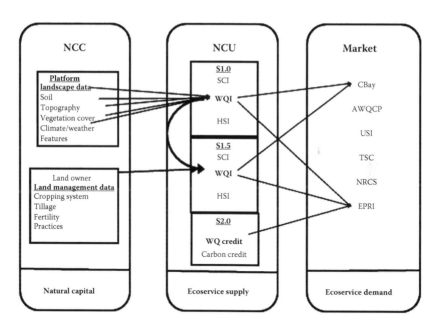

Figure 9.6 Symbiotic demand is defined as mutually interdependent transactions between a single land manager and multiple stakeholders procuring the same unit of ecoservices (NCU). The transaction is an S1.0 exchange between the platform managers and each demander (CBay and EPRI), and an S1.5 exchange between the same parties. Costs of the S1.0 and S1.5 are potentially shared between the three parties for a single NCU as club goods. The second transaction set is depicted as a S2.0 exchange between the land manager and EPRI as a private good. The arrows from the NCC section depict the flow of data to generate S1.0 and S1.5 NCUs. Arrows from the NCU to the Market sections depict trades.

In this scenario, CBay and EPRI purchased S1.0 and S1.5 data and EPRI also purchased S2.0 credits. The first transaction set consists of an S1.0 exchange between CBay and the platform managers, and an S1.5 exchange between CBay and the land manager. The second transaction set depicted is a S1.0 exchange between EPRI and the platform managers and S1.5 and S2.0 exchanges between the land manager and EPRI. Since NCU values generated from each NCC may have relatively low economic value, a cryptocurrency or e-currency system (discussed in Chapter 11) embedded with the MSSG platform could make small exchanges feasible and also reduce transaction costs related to monetary exchanges.

Only two transaction paths are depicted in Figure 9.6 to reduce the graphic clutter, but in a functioning MSSG platform S1.0 and S1.5 values could be part of transactions with each of the sustainability demanders as club goods. The S2.0 value, as a private good, could only be exchanged with one demander: EPRI in this case.

This symbiotic demand scenario fosters a shared governance approach as each organization contributes interdependently as they deem appropriate at the point of service. Shared governance is applicable to the emerging business *ecosystem* concept that rejects the idea of a "single win" in traditional enterprise competition. The business ecosystem theory emphasizes that the business environment is a closely linked and mutually dependent symbiotic system, and enterprises need to work with others to develop solutions (Liu et al., 2013).

section three

Designing a glocal business ecosystem

chapter ten

Governance of the glocal commons

The *commons* are the cultural and natural resources accessible to all members of a society, including the basics for life such as air, water, and a habitable earth that are held in common, not owned privately. Since people identify with the *commons* from many perspectives, the commons are often defined differently or open to interpretation. The International Association for the Study of the Commons (IASC) broadly defines commons as any natural or man-made resource that is or could be held and used in common (Berge and Laerhoven, 2011).

The term *glocal commons* is used to describe how the commons are viewed in an increasingly interconnected world. González-Gaudiano (1997) described *glocal* as an integrated dimension of the global and the local affecting each other reciprocally. This term provides a more inclusive frame of interpretation of global environmental issues such as climatic change, the loss of biodiversity, and desertification by analyzing global environmental issues within the context of their effect on local conditions and the local actions needed to address them.

The challenges of governing local commons were described and made famous by Hardin's (1968) "Tragedy of the Commons" essay. Today, similar scenarios are played out throughout the world. In an interconnected and glocalized economy, governance of the local commons is in many respects governance of the global commons.

10.1 Glocalization phenomenon

Generically speaking, glocalization affects a variety of issues. Glocalization occurs when local actors and actions have a more pronounced role in or effect on global issues. It denotes a merging of global opportunities and local interests to create a more socioeconomically integrated world (Roldan, 2011). It is the result of the combination or convergence of global and local processes and interactions (Vries, 2010; Wellman, 2002). The term *glocalization*, originated from the Japanese word *dochakuka*, which simply means *global localization* and referred to a way of adapting global farming techniques to local conditions. The term evolved into a marketing strategy when Japanese businessmen adopted it in the 1980s (Popova, 2009).

The Canadian sociologist Barry Wellman began using the term in the 1990s to refer to people who are actively involved in both local and wider-ranging activities of friendship, kinship, and commerce. Wellman describes glocalization as a transitional stage between local groups and broad networks. This transition was driven by revolutionary developments in both transportation and communication. It was a move away from small isolated communities to creating connections between people in different places and multiple social networks (Wellman, 2002). The term *glocalization* carries a broader perspective of the commons as it points to the *simultaneity* of globalizing and localizing processes, and to the interconnectedness of the global and local levels (Blatter, 2007).

10.1.1 Social effect

Glocalization is primarily a social effect. Wellman (2002) identifies three social groups on a glocalization continuum referred to here as local groups, glocal clusters, and networked individuals (Figure 10.1). They are represented as three separate types, but the groups often contain characteristic or mixtures of the three.

Wellman (2002) describes this social evolution beginning with local groups in a single location. The inability to communicate and travel with ease favors a singular place-based structure. Because of limited outside influence, the community and its commons are generally well-defined physical entities.

The second phase occurred as communication and transportation networks expanded. In glocal clusters, people redefined their community based on interests, work, and family values, rather than just the characteristics of the immediate surroundings. Interactions occur among people in multiple places, but are not truly mobile.

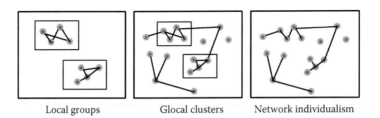

Local groups   Glocal clusters   Network individualism

Figure 10.1 Gocalization occurred as revolutionary developments in communication and transportation systems allowed greater social interaction. Local groups remained isolated with little outside communication and travel. Glocal clusters formed as people interacted but the individuals were still place-based. Mobility in communication systems created networked individuals. (From Wellman, B., *Digital Cities II: Computational and Sociological Approaches,* Springer, Berlin, 2002.)

The third evolutionary phase is networked individualism, where connections are person-to-person and occur in a highly mobile society where place becomes secondary. Communities and commons may be defined by interconnected individuals with very specific values. People are now able to experience the global commons personally as communication and transportation can bring individuals nearly anywhere.

Each of these social phases has influence in the agriculture production system as it *is*. Local groups are those in agriculture relatively isolated from various trends in glocal food, social, and environment trends, but deeply connected to local cultures. Glocal clusters represent a large portion of the agriculture community today beginning to interact with social, corporate, and environmental groups. The most recent phase of network individualism has influenced how globalized food corporations view their supply from a more glocal perspective and personalized their brands. Due to the place-based nature of agriculture, farmers and land managers cannot transition to the truly mobile networked individual, but they and the food processers are responding to their networked individualized consumers.

10.1.2   *Sustainability supply chain effect*

Glocalization of the world's markets results in an increased mobility of goods, services, labor, technology, and capital throughout the world (Roldan, 2011). In agribusiness, global firms seek the "sweet spot" to be found by effectively marrying *globalization* to *localization*. Nestlé and the Coca-Cola Company are embracing glocal-type concepts. Nestlé views food as a local matter and attempts to "centralize what you must, but decentralize what you can." Coca-Cola Company's strategy is to "mingle global and local" by utilizing local suppliers and local bottlers, employing local people, and addressing local culture and taste (Kelly, 2015).

In supply chain sustainability, glocalization highlights the need to address ecosystem impacts and the *multiple commons* of a supply chain. For example, carbon sequestration impacts ecosystems glocally: it improves soil health at the farm scale, it improves water quality at the regional watershed scale, and it reduces atmospheric carbon and climate change potential at the global scale—demonstrating the *simultaneity* of globalizing and localizing processes (Blatter, 2007). A glocalized perspective reveals the economic and ecological complexity and their relationship to the commons.

To address these sustainability challenges means to think and act at multiple levels of governance simultaneously and to be able to connect the *global* and the *local* (Young et al., 2014). Corporate supply chain efforts, such as The Sustainability Consortium (TSC), are attempting to account for these glocal values provided by the sustainable management of soil, water, and air resources throughout their supply chain. While TSC and others rely on terms such as *sustainability* and *the environment* rather than on the commons, one could suggest these have become the modern terms for the commons and addressing their complete supply chain is analogous to addressing the *glocal commons* impacted by their economic activities.

10.1.3   Local + global = glocal commons

The glocalization effects on society and markets are primary influences on how the perspective of the commons has evolved. In the nineteenth century much of the world lived in what Wellman (2002) described as little boxes or local groups with a place-based mentality. The commons were viewed in the manner William Forster Lloyd wrote about in an 1833 pamphlet. His observations were of a degraded pastoral commons populated with thin cattle as compared to privately owned pastures that had healthier cattle (Lloyd, 1833). Lloyd stated that the inequality in production did not lie in natural or acquired fertility, but how the land and cattle were managed. He noted that those that owned the land and cattle were more apt to properly manage the lands by removing cattle as pastures became degraded. His astute observations defined and described the commons from a place-based scenario within the context of a local economy and recognized the limited (natural capital) capacity of the pasture.

In 1968, Garrett Hardin, inspired in part by Lloyd's pamphlet, wrote the *Tragedy of the Commons* essay, not just to discuss the perils of livestock grazing in the commons, but to use this as a global analogy to state that, "as long as we behave only as independent, rational, free-enterprisers we are locked into a system of ruin and fouling our own nest" (Hardin, 1968). Hardin agreed with Lloyd's conclusion, stating property rights and the self-interests of individuals could avert the "tragedy of the commons" as it relates to land-based production. But, he added, those same self-interests favor pollution as the economical rational man discharges wastes into the commons associated with water and air (Hardin, 1968). Since the waters and air could not be owned, unlike the landscape's pastures, Hardin suggested government coercion through regulations is an option. Hardin recognized the commons beyond the pasture to include those resources both directly and indirectly impacted by misuse or lack of good governance of the landscape.

This broader, regional vision of the commons became apparent as forests died due to acid rain, rivers burned, and smog enveloped cities. In the latter part of the twentieth century, governments began recognizing the human effect on the global commons related to the ozone layer, the climate system, the biosphere, and the hydrosphere. It became a concern that if the planet exceeded a saturation point for these commons, it may undermine social and economic development (UNEP, 2011). As the acceptance of human dependence on the glocal commons grows so does the motivation for governments, utilities, nongovernmental organizations (NGOs), and corporations to account for the impacts and find pathways to develop governance strategies to address them.

In reference to the Rio+20 Conference Sustainability Development Goals (SDG), Young et al. (2014) state that there is no way to secure a sustainable future in which both biophysical and socioeconomic processes are tightly interwoven without connecting the global and the local. To do so, a strategic pathway is needed for stakeholders to think and act within multiple levels of governance simultaneously.

10.2 Glocal governance pathways

The pathways to governance strategies begin from the perspective and culture of the organizations seeking new models. Gupta et al.'s (2013) research on water governance history revealed that water governance, like agriculture sustainability governance, consists of a complicated and uncoordinated set of governance strategies. Gupta (2007) identified the three pathways organizations take to develop governance strategies: top-down, bottom-up, and a private sector approach. She states that none of the paths has the monopoly as the best approach to develop a glocal governance strategy, and they all, to some extent, do not meet their expected goals.

The bottom-up approach may exclude marginal groups if the rules are made by the most powerful at the cost of others. The top-down pathway, while seemingly scientifically motivated, has political overtones and often neglects the importance of local context. The private sector pathway, driven by the capitalist and mercantilist ideology, may overlook social and environmental consequences. These pathways of bottom-up, top-down, and private sector often lead to the familiar governance traits and styles of network, hierarchy, and market governance, respectively. A fourth pathway toward adaptive governance has the potential and characteristics of shared governance.

10.2.1 Bottom-up sustainability governance

A bottom-up approach is analogous with governance at the level of Wellman's (2002) local clusters where individuals network to address legal and social issues as they emerge (Gupta, 2007). Ostrom's (2011) research on local groups with long-enduring common pool resource governance models revealed that bottom-up governance could be successful when resource boundaries were clearly defined, rules addressed local conditions, participants could change rules and monitor the resource, conflicts were fairly resolved, and resource users have long-term rights to the resource.

The use of bottom-up governance to address water can be traced back 6000 years as it was among the first resources that humans used and tried to manage, and hence one of the first areas in which social rules and customs have been developed. Historically, it emerged as a slow buildup of precedent-setting concepts based on existing customs and practices (Gupta, 2007).

Of the 11 governance case studies used in this text, two could be considered bottom-up. The United Suppliers SUSTAIN project was initiated by private practitioners directly involved in managing landscapes. They used a network-lead governance approach. The Agricultural Environmental Quality Assurance (AgEQA) project was initiated by the Chisago Soil and Water Conservation District (SWCD), a public practitioner agency, and used a network-participant governance approach. In many respects, both were initiated by emerging top-down demands, but organized from the bottom up to address those demands from their own perspective.

10.2.2   Top-down sustainability governance

Policy makers in top-down environmental governance often rely on the expertise of scientists and engineers to determine the best scientific way to manage resources. The communication and transportation channels of Wellman's (2002) glocal clusters aid in maintaining interaction and delivering a consistent message to various social sectors. Governments are often in the best position to connect with local groups to support the top-down hierarchy environmental governance by government.

Those concerned about loss of biodiversity, soil erosion, desertification, deforestation, decline of fisheries, and other environmental issues relied on government and hierarchy governance to address market failures (Lemos and Agrawal, 2006). This top-down, state-based governance remained the primary means to address environmental concerns until as recently as the early 1980s. According to its critics, this approach to environmental policy is costly, inflexible, and often unresponsive to the needs of particular communities and citizens. These failings are attributed to a lack of efficiency, knowledge, and accountability on the part of government, and the excessive influence of special interests, including business and environmental lobby groups, who impose their own narrow agendas on the policy process (Oddie, 2004).

These concerns, among others, have contributed to a growing weakness in governments as it relates to glocal environmental governance. Claussen (2001) concluded that the United States and other countries will not be able to deal effectively with environmental issues of the commons such as global climate change, biodiversity loss, and marine conservation without a better system of environmental governance. She concluded that the private sector needs to respond to address the concerns of society in a manner that reduces costs, increases flexibility, and provides leadership in (glocal) environmental issues. If governance of the global environment is allowed to remain in the exclusive or near-exclusive realm of governments and institutions, the global environment will be poorly served.

Of the 11 case studies, five (SCS, NRCS, LEQA, AWQCP, and CBay) followed a government-centric, top-down pathway. All five, except for the Livestock Environmental Quality Assurance (LEQA) program using network-lead governance, instituted a hierarchy governance strategy.

10.2.3   Private sector sustainability governance

The third pathway is based on the forces of globalization within the context of capitalism in the private sector (Gupta, 2007). Lemos and Agrawal (2006) see the most important emerging trends shaping environmental governance as globalization, decentralized governance, market and individual-focused instruments, and cross-scale governance.

These glocal forces are creating shifts in environmental governance from state-based hierarchies toward market and network governance styles. The rise of private governance signifies a new phase in the ongoing process to establish environmental standards with an evolution in the direction of market-oriented, deregulatory systems of governance (Falkner, 2003). In many respects, it is the glocalization effect: the increased interconnections due to unlimited communication and ease of physical transportation enabling cooperation of many different actors across local, regional, national, and global levels within the economic, political, social, and cultural domains.

The motivation to enter this pathway is often initiated by the private sector seeking new ways to address government regulations and responding to social and consumer demand of more sustainable supply chains. Of the 11 case studies, four (MMEQA, TSC, FtoM, and EPRI) are considered private sector efforts. The Electric Power Research Institute, Inc. (EPRI) was initiated to address government (TMDL) regulations, the Minnesota Milk Environmental Quality Assurance (MMEQA) program was initiated to address regulations and corporate supply chain concerns, and TSC and FtoM were initiated to address sustainable supply chains.

On this governance path, several related, but separate options include governance through public–private partnerships and more strictly defined private governance strategies such as second-order agreements and adopting sustainable supply chain procurement standards.

10.2.3.1   Public–private partnerships

Over the last 15 years, the number of global public–private partnerships (PPPs) for the environment has grown exponentially (Andonova, 2010). In the context of U.S. environmental policy, public–private partnerships appeared two decades ago when the private sector provided state and local wastewater treatment services in collaboration with the U.S. Environmental Protection Agency (EPA) (Bhan, 2013). PPPs occupy a middle ground in the increasingly complex continuum between public and private environmental governance (Andonova, 2010). PPPs are distinct from private governance in that they involve the deliberate pooling of authority, competences, and resources from both the public and private spheres.

Several definitions of PPPs are used. Hodge and Greve (2005) define PPPs as cooperative institutional arrangements between public and private sector actors. A more detailed definition is an agreement for collaborative governance between governments and nonstate actors such as foundations, firms, and advocacy organizations to establish common norms, rules, objectives, and decision making and implementation procedures for a set of policy problems (Andonova, 2010). Some scholars view PPPs as privatization of public service delivery, while others argue that they are a new governance tool (Bhan, 2013).

In PPPs, the governance authority is negotiated, rather than granted through delegation, market mechanisms, or moral recognition and is subject to contestation and renegotiation. Increasing enthusiasm for PPPs around the world is built on the premise that public policy problems— addressed exclusively by the public sector in the past—are becoming more complex and need to be solved using a collaborative endeavor that takes advantage of the specialization and professional expertise offered by the private, semi-private, and nonprofit sectors. It is third-generation policy problems, such as climate change and sustainability, considered to be classic wicked problems that require cross-sector expertise for long-term problem solving (Bhan, 2013).

Not all environmental issues need a PPP strategy or even demand public sector involvement. In these cases, a narrower perspective of private governance emerges to address sustainability issues within corporate supply chain boundaries or to address government regulations with private actors.

10.2.3.2 Solely private sector

Private governance in environment protection has been increasing at the level of individual firms, industries, and cross-sector organizations (Falkner, 2003). Over the past two decades, the role of private parties in environmental regulation has grown dramatically (Stafford, 2012). The emergence of private environmental governance suggests this different approach to managing common pool resources and reducing environmental externalities is possible (Vandenbergh, 2013). A number of companies noted that sustainability is increasing in business importance in such areas as product design, procurement, and collaboration with external stakeholders and that sustainability was tied to a company's business success (Feldman, 2009).

Private systems of governance are typically defined as those established and managed primarily by nonstate actors. They can derive their governance authority from their capacity to introduce environmental information in the chain of production in specific markets and from the moral authority and norms promoted by advocacy organizations (Andonova, 2010). Private governance is more extensive than just cooperation between private actors to achieve mutually beneficial objectives; it becomes institutionalized and has a permanent nature (Falkner, 2003).

The rise of private actors in environmental governance coincides with an ongoing erosion of state capacity (Claussen, 2001) and an increase in relationships between firms, states, and civil society (Falkner, 2003). Private governance has become a reality in environmental politics that few analysts deny and this recognition is changing the traditional assumption that government is the only actor and regulation is the only instrument for addressing environmental problems.

Not all environmental problems are suitable for private governance initiatives, but new forms have emerged that are worthy of the attention of policy makers and practitioners, particularly in light of the national and international gridlock on major environmental problems (Vandenbergh, 2014). Two prominent issues being addressed by private governance are *second-order agreements* and *sustainability supply chain standards*.

10.2.3.2.1 Second-order agreements Second-order agreements are created between private entities to address environmental issues in response to the existence of (first-order) government regulations or to address issues not regulated but deemed important by the public or the organizations themselves (Vandenbergh, 2005). Private entities enter into contractual agreements to sort out the roles and responsibilities to address environmental regulations or address risk management issues that lie outside the interest or reach of government agencies. The traditional view of environmental regulation being an adversarial system that pits regulated private entities against a public regulatory agency does not provide an active role for private parties. In reality, the relationship between private regulated entities is much more complex (Stafford, 2012).

When addressing government regulations, second-order agreements reveal who actually pays the costs of regulatory requirements and thus who has incentives to develop, implement, and enforce regulatory requirements. This is revealed through standard development, certification verification, labeling, and the overall system function (Vandenbergh, 2013). Second-order agreements may improve regulatory quality by inducing more efficient implementation of public regulations. In short, second-order private agreements go to the heart of the accountability and efficacy of public regulation (Vandenbergh, 2005).

In the absence of government regulatory requirements, second-order agreements are a legal vehicle by which public preferences can be channeled by nonprofit groups into regulatory requirements on private firms, bypassing agencies and elected officials (Vandenbergh, 2005). A far-reaching private initiative that affects environmental performance is the ISO 14001 certification program by the International Organization for Standards, an international nongovernmental organization. The ISO 14001 certification program essentially works as a labeling system, conveying information to potential investors and consumers about the environmental standards to which certified companies adhere (Stafford, 2012).

Vandenbergh (2013) sees the potential for private environmental governance to become a coherent, discrete concept that affects environmental behaviors and public governance by offering new solutions to environmental problems. A well-known example of a U.S. industry-led initiative is the Responsible Care Program, introduced by the Chemical Manufacturer's Association in 1988 partly in response to the acute Bhopal disaster (Stafford, 2012). More recent examples of private governance organizations emerging from more chronic natural resource issues are the Forest Stewardship Council (FSC) and the Marine Stewardship Council (MSC), both of which seek to manage natural resources by setting supply chain and procurement standards.

10.2.3.2.2 Supply chain and procurement standards   Private environmental governance emerges when private entities take actions to manage the exploitation of common pool resources. Private certification systems regulate 14% of the temperate forests and 7% of the fisheries around the world. The extent of private governance is increasing. Corporations spend more than $500 million on environmental investigations in connection with commercial transactions annually, greater than the $400 million annual budget of the entire EPA enforcement office. Private firms routinely impose environmental requirements on their suppliers. In some sectors, private corporate supply chain requirements are becoming the de facto constraints not because of government requirements but because their supply contracts require them to comply with a private standard (Vandenbergh, 2013).

Corporate supply chain requirements are driving substantial amounts of carbon emissions reductions without regard to international boundaries, and corporate policies by companies such as Walmart and Target are becoming the de facto regulatory floor product sustainability. These environmental requirements are not just a form of market behavior, but a new form of governance (Vandenbergh, 2014). He states that a more inclusive governance strategy will occur if public officials begin to recognize these private sector transactions are a form of private environmental governance affecting public environmental laws.

10.2.4   Adaptive pathway

The governance pathways (bottom-up, top-down, and private sector) taken by organizations are often based on their culture or the need to address a particular problem. Often, organizations end up adopting governance structures based on their successes in addressing single sector issue (Scholz and Stiftel, 2005). Since no organization can have a complete perspective of glocal agriculture sustainability issues, these strategies fall short in acquiring the information and engaging the breadth of stakeholders required for generating long-term sustainable solutions.

The grouping of six case studies in Figure 9.5 is an example of common efforts originating from different governance perspectives, adopting different governance styles and acquiring disparate data sets. An adaptive governance strategy, on the other hand, supports coordination of independent stakeholders, authorities, knowledge, and interest groups: a necessity in addressing wicked problems (Scholz and Stiftel, 2005).

Chaffin et al. (2014) define adaptive governance as a range of interactions between actors, networks, organizations, and institutions in pursuit of a desired state for socioecological systems. Hatfield-Dodds et al. (2007) state that the concept of adaptive governance emerged from the self-governance research by Elinor Ostrom and defines it as an evolution of formal and informal institutions for the management and use of shared assets, such as common pool natural resources and environmental assets that provide ecosystem services.

Termeer et al. (2010) state that the field of adaptive governance has the ambitious goal of developing new governance concepts that can handle the inherent complexity and unpredictability of dynamic socioecological systems. It must address a world that is characterized by both continuous and abrupt changes, often with largely unpredictable consequences.

10.3   Glocal common governance

It is fairly straightforward as to why a variety of governance strategies emerge from the multiple sectors and organizations within the system as it *is*. The glocalization of agriculture sustainability presents many visions of what sustainability means at various scales and as a result, organizations acting within their relatively narrow scope and culture are apt to produce a variety of programs and governance strategies. Despite this leading to conflicts and inefficiencies, organizations seemingly have few choices and will continue to make their best attempts to address their agriculture landscape objectives.

Ideally, a system of adaptive governance emerges and connects actors and institutions at multiple levels enabling ecosystem stewardship (SRC, 2015). Ostrom's (2011) work on self-governing institutions has identified numerous examples of locally evolved autonomous institutional arrangements that have sustained common pool resources for centuries. Her work has challenged the presumption that the tragedy of the commons is the inevitable result or that it can only be averted or resolved through externally imposed expert management, such as government regulations.

10.3.1   Local commons

In *Struggle to Govern the Commons*, Dietz (2003) identified general principles to create robust local governance institutions using both formal and informal social entities. When social entities are formal, they are commonly identified as institutions, such as the U.S. Congress, the Roman Catholic Church, or academic institutions. Informally, institutions are the less tangible components of social order and organization, reflecting culture, norms, and customs. Informal institutions define the roles and responsibilities of resource users (Rahman et al., 2012) and are often invisible to government and society as they exist as shared concepts in the minds and routines of participants (Ostrom, 2007). Olsson et al. (2006) described shadow networks as informal connections with unique social tools and the willingness to experiment and resolve problems.

Adaptive governance within local institutions enables participants to acquire the necessary information, deal with conflict, induce compliance, and provide technological infrastructure. Dietz (2003) identified characteristics of these self-governed systems:

- Align rules with ecological conditions
- Clearly define boundaries for resources and user groups
- Devise accounting mechanisms for monitors of the resource
- Use graduated sanctions for violations
- Establish low-cost methods for conflict resolution

These general principles for robust governance institutions for localized resources are well established as a result of multiple studies. Many of these also appear to be applicable to regional and global resources, although they are less well tested at those scales (Dietz, 2003).

10.3.2   Glocal commons

Developing adaptive governance systems to address agriculture landscape sustainability and other issues of the global and glocal commons will require sustained research and an understanding of national and international policies. Such experiments can yield the scientific knowledge necessary to design appropriate adaptive institutions. Dietz (2003) offers three more self-governing principles to apply established governance concepts at regional and global scales:

- Analytic deliberation: Dialog involving scientists, resource users, and interested publics provides improved information and the trust in it that is essential for information to be used effectively, builds social capital, and can allow for change and deal with inevitable conflicts well enough to produce consensus on governance rules.
- Nesting: Institutional arrangements must be complex, redundant, and nested in many layers.
- Institutional variety: Governance should employ mixtures of institutional types (e.g., hierarchies, markets, and networks) that employ a variety of decision rules to change incentives, increase information, monitor use, and induce compliance.

Resolving problems at this scale inevitably involves complex collective action spanning jurisdictions and generations, requiring new governance mechanisms (Esty and Ivanova, 2002). One strategy to govern the glocal commons is to re-create the characteristics of the successful, local self-governance system identified by Dietz (2003) at glocal scales. Esty and Ivanova (2002) propose the creation of a global environmental mechanism that draws on information age technologies and networks to promote cooperation in a lighter, faster, more modern, and effective manner than traditional institutions.

Successful governance strategies for the glocal commons must provide disparate stakeholders with a platform to manage the varied scope and scale of natural capital outputs and reduce the inherent conflicts of governance styles. The challenge is how to combine different governance approaches successfully "on the ground" in a tailor-made way that is reflexive and dynamic and at the same contributes to common, universal goals (Meuleman and Niestroy, 2015).

Scharmer and Kaufer (2013) ecocreative ecosystem creates new space for cross-sector innovation and engagement and is aligned with the use of a business ecosystem approach: a business model with the capacity to process the new complexities of glocal commons accounting, valuation, and commerce.

chapter eleven

Designing a business ecosystem

Sustaining agricultural landscapes and governing the glocal commons remains an unresolved wicked problem. The diverse and disparate stakeholders, conflicting governance styles, and challenges of accounting for natural capital ecoservices is currently overwhelming for the corporate, government, and nongovernmental organization (NGO) entities and sectors to address.

In Deloitte's *Business Ecosystems Come of Age*, Eggers and Muoio (2015) see a trend by which many kinds of wicked problems are being recast as *wicked opportunities* and resolved though *solution ecosystems*. Unprecedented networks of nongovernment organizations, social entrepreneurs, governments, and businesses are coalescing around seemingly unresolvable socioeconomic issues. Moore (2006) referred to this opportunity space as a *business ecosystem*.

Thomas and Autio (2013) define a business ecosystem as a network of interconnected organizations around a focal firm or a platform incorporating both production and use side participants with a focus of cocreation of new value through innovation. Moore (1993) described a *business ecosystem* as a space where leaders co-envision and comanage coevolution among members. Leaders in a business ecosystem establish what might be called community governance to achieve *collective action* in a manner similar to democratic and quasidemocratic communities (Moore, 2006). The concept and function of business ecosystems are believed to be capable of better explaining how multisided businesses evolve (Baghbadorani and Harandi, 2012).

11.1   Evolving ecosystems

Ecosystem is a term first coined in 1935 by British botanist Arthur Tansley to describe a community of living organisms and nonliving components (air, water, and mineral soil) linked together through nutrient cycles and energy flows. Types of ecosystems are defined by the network of interactions among organisms, and between organisms and their environment. Moore (1993) expanded the use of the term by suggesting that a company be viewed not as a member of a single industry but as part of a *business ecosystem*.

11.1.1   Business ecosystems

The business ecosystem is a new and important stream of theory in the field of strategic management. It uses metaphors and concepts from ecology and develops a new way of looking at relations between firms. Businesses are seen as interconnected and interdependent members of *ecosystems* that coevolve and share a common fate (Baghbadorani and Harandi, 2012).

In a business ecosystem, companies coevolve capabilities around new innovations by working cooperatively and competitively to support products, satisfy customer needs, and eventually incorporate the next round of innovations (Moore, 1993). A business ecosystem can also be conceived as a network of interdependent niches that are more or less open to the world of potential contributions and creative participants (Moore, 2006). Business ecosystems are dynamic and coevolving communities of diverse actors typically bringing together multiple players of different types and sizes in order to create, scale, and serve markets in ways that are beyond the capacity of any single organization. Competition, while still essential, is certainly not the sole driver of sustained success. Participants are incentivized by shared interests, goals, and values, as well as by the growing need to collaborate in order to meet increasing customer demands, from which all can derive mutual benefit (Kelly, 2015).

In contrast to the conventional value chain view, business ecosystems offer a dynamic, systems view. This view consists of the value chain of a business and those with rather indirect roles, such as companies from other industries producing complementary products or equipment, outsourcing companies, regulatory agencies, financial institutes, research institutes, media, universities, and even competitors. Economic competition no longer resides at the business level, but competition is formed and defined between business ecosystems (Baghbadorani and Harandi, 2012).

Business ecosystems do differ from natural ecosystems, in that the actors in business ecosystems are intelligent and are capable of planning and picturing the future with some accuracy. Second, business ecosystems compete over possible members. This kind of behavior is not observed in nature. Third, business ecosystems are aiming at delivering innovations, where natural ecosystems are aiming at pure survival (Liu, 2013). With this conscious guidance, an increase in the rate of ecosystem evolution is expected as the ecosystem concept continues to become more embedded in business culture and processes.

11.1.2   E-commerce ecosystems

Amor (2000) credits International Business Machines (IBM) for coining the term e-business in 1996 to describe the use of information and communication technology to enable relationships with individuals, groups, and other businesses via the Internet. The concept of a digital business ecosystem emerged in 2002 by adding *digital* to Moore's (1993) *business ecosystem* (Nachira et al., 2007).

Lihua et al. (2010) defined e-business ecosystems as an organic ecosystem that is made of enterprises and organizations with close relations, using the Internet as a platform to create alliances, share resources, and make full use of their advantages beyond geographic limits. The basis for this definition was the Alibaba Group, an e-retailer whose business model and prospectus is based on the ecosystem business model. The term *e-commerce ecosystem* is analogous to e-business and digital business. It is a term used by Liu (2013) to describe the application of natural ecosystem theory to business ecosystems. Liu (2013) traces this evolution back to the study of systems theory, to the application of ecological theories and business methods to reach this advanced stage of e-commerce ecosystems.

11.1.3 An eco-commerce ecosystem concept

An eco-commerce ecosystem is described as the next conceptual phase on the continuum of natural and business ecosystem evolution. Figure 11.1 depicts this evolution beginning with the natural ecosystems that humans rely on for food, shelter, and other provisions. As society evolved, businesses were created and their interactions eventually evolved into business ecosystems (Moore, 1993).

With the advent of the Internet, big data, and communication networks, business ecosystems evolved into increasingly connected e-commerce ecosystems. The Alibaba Group, the largest online and mobile

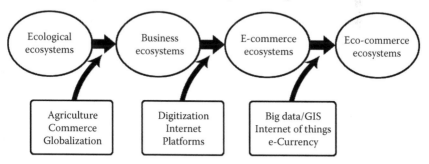

Figure 11.1 Ecosystem *types* are in a constant state of evolution beginning with early humans interacting directly with ecological ecosystems for all their provisions. Agriculture and commerce led to the development of business ecosystems. With the advent of digitization, the Internet and platforms, e-commerce ecosystems emerged in the 1990s. With big data, GIS, algorithms, and e-currency systems, ecological benefits can be accounted for and exchanged in eco-commerce ecosystems. (Adapted from Liu, H., Z. Tian, and X. Guan. *International Journal of U- and E-Service, Science and Technology* 6, no. 6, 41–50, 2013.)

commerce company in the world in terms of gross merchandise volume, operates an ecosystem using a multisided platform for third parties. In its prospectus, it uses the term *ecosystem* 160 times.

It is through the knowledge gained from the ecological, business, and e-commerce ecosystems that an eco-commerce ecosystem can be designed. Incorporating geographical information system (GIS) capabilities, natural capital units (NCUs), and the capacity to conduct transactions are new traits in the evolution toward an eco-commerce ecosystem. In a nutshell, ecological ecosystems beget business ecosystems, which beget e-commerce ecosystems, which beget eco-commerce ecosystems.

The eco-commerce ecosystem concept has capacities similar to Esty and Ivanova's (2002) global environmental mechanism that would characterize environmental problems, create space for environmental negotiation, and sustain a buildup of capacity to address sustainability issues. The eco-commerce ecosystem contains a negotiation space where ecoservice values of the natural ecosystem interface with the ecoservice capacities of the e-commerce ecosystem. In other words, the functions and processes of the natural ecosystem are accounted for by using the functions and processes of the e-commerce ecosystem. Resolving the complexity issues of the intertwining ecosystems is the function of an eco-commerce ecosystem.

11.1.3.1 *Interface of ecoservice values*

The plane or point where the ecosystems interface is where complexity converges, values are accounted for, and sustainability issues can potentially be resolved. An interesting, yet somewhat obvious revelation at this interface is the recognition of the similarities of the definitions of e-commerce ecosystem services and ecological ecosystem services.

The Alibaba Group states that it does not engage in direct sales, compete with their merchants, or hold inventory, but provides the fundamental technology infrastructure to help businesses leverage their capacities to conduct commerce. In other words, the value of the Alibaba Group resides in providing the *structure, processes, and functions* of the e-commerce ecosystem to enable other businesses to create value within the Alibaba Group ecosystem. These structures, processes, and functions are the *ecoservices* of the Alibaba ecosystem. The entire valuation proposition of the Alibaba Group is based on creating these ecoservices.

Perhaps not just coincidently, the wicked challenge of sustaining the agriculture landscape is how to value the structure, processes, and functions (ecoservices) of the agroecosystems. It is at the interface of the natural ecosystems and the business ecosystems that new ecoservice valuation strategies lie. The concept of the eco-commerce ecosystem is based on the application of the structure, processes, and functions of an e-commerce ecosystem to account for the ecoservices (structure, processes, and functions) of the agricultural landscape ecosystem. The parallels between natural and business ecosystems identified by Moore (1996) continue to unfold as greater complexity is revealed.

11.2 Ecosystem design layers

Designing a platform-based business ecosystem is unlike a traditional business supply chain where a product is produced and then delivered into an existing or developing market. When designing an ecosystem, it is the interaction of the participants and their roles, not the market, that is essential (Adner, 2006).

Design considerations for a business ecosystem include the environmental layer, participant layers, and risk management. The environmental layer includes the economic, technology, natural, social, cultural, law, policy, and financial groups and issues that surround the participants and frame the entire ecosystem (Yu et al., 2011). The participant layer includes the leaders, users, and contributors: those that create, support, and interact with the platform. The risk management layer includes typical business assessments as well as issues associated with users and contributors of the platform that are not under direct control of the platform builders.

11.2.1 Participant layers

Platform participants can be described in multiple ways. Yu et al. (2011) identified three internal groups (*leader, key,* and *support* populations) and two external groups (*related* and *parasitic* populations) to describe the populations in a business ecosystem. Baghbadorani and Harandi's (2012) similar model is used here and describes three participants layers: leaders, users, and contributors.

11.2.1.1 Leader layer

The leaders' layer contains the platform, vision and governance (Baghbadorani and Harandi, 2012) of the business ecosystem. Lihua et al. (2010) identifies this layer as the core of the business ecosystem, providing the platform and regulatory services that integrate and coordinate the ecosystem. It sets the vision and standards for the other members of the ecosystem to follow. The main value that the business ecosystem leader brings to an ecosystem is the platform and the tools and frameworks it provides.

11.2.1.1.1 Platform Creating an ecosystem begins with building the platform to create the conditions that enables interaction. Since value is produced by the users, the platform must overcome obstacles such as the chicken-and-egg dilemma, converting consumers into producers, encouraging producers to produce more, and providing an adequate incentive for users to engage and interact on the platform (Choudary, 2015).

The platform may be monetized through strategies such as a transaction fee, a producer fee for access, revenue from advertisements, or a freemium strategy where enrollment is generally free with users paying for more advanced tools. Participants will support the monetization strategy if it brings value to the participants. Hagel (2015) categorized multisided platforms into types based on what they do for their participants:

1. Aggregation platforms focus on transactions or tasks to be completed. These tend to operate as a hub-and-spoke mode supporting generally short-lived relationships based only on a particular transaction. Transactions are usually brokered by the platform owner and organizer. These platforms aggregate data, participants, and collaborators.
2. Social platforms are similar to aggregation platforms in that they aggregate people, but focus on building long-term relationships rather than just a transaction. The connections are network-based, rather than a hub-and-spoke model and so transactions or communications do not usually go through the platform owner.
3. Mobilization platforms are designed to connect a group of people to accomplish something beyond the capabilities of any individual. Because of the need for collaboration, these platforms tend to foster long-term relationships to achieve a shared goal. Examples include open source software platforms and social support platform movements.
4. Learning platforms have an emphasis on small work groups with trust-based relationships that take on a particular challenge. These platforms not only create efficiencies in collaboration and transactions as the other platform types, but also grow the participants' knowledge and accelerate performance improvement.

Hagel (2015) sees the first three platforms listed as having the potential to evolve into learning platforms. Businesses that find ways to design and deploy learning platforms will likely be in the best position to create and capture economic value in an increasingly challenging and rapidly evolving business environment.

11.2.1.2　User layer

The user layer contains the key species of the ecosystem: the customers involved in business trading, as well as consumers, retailers, manufacturers, and suppliers (Lihua et al., 2010). The users are those that interact directly with the core functions of the platform (Yu et al., 2011). Users are the vital components as they purchase the products and services that business ecosystems are formed to produce. Hence, without users, formation of an ecosystem could be meaningless.

11.2.1.3   Contributor layer

The contributor layer includes supportive and parasitic species (Lihua et al., 2010) consisting of numerous interdependent organizations and individuals contributing to the evolution of a business ecosystem. These organizations actively work on platforms to improve their performance, while extending the capabilities of the platform. Contributors carry out tasks related to design, production, operations, distribution and delivery of products, solutions, and services (Baghbadorani and Harandi 2012). Considered supportive species, they are not dependent on the business ecosystem for their survival and include logistics companies, financial institutions, telecommunication suppliers, and government agencies.

Parasitic species are companies that must coexist with the business ecosystem. They rely on the ecosystem for survival. Typical instances are value-added service providers, such as technology vendors, advertising and marketing service providers, training agencies, and consulting firms (Lihua et al., 2010). These businesses and the business ecosystem rise or fall as one (Yu et al., 2011).

11.2.2   Risk management layer

Business ecosystems contain greater uncertainties than a typical supply chain due to the dynamic nature and relationships associated with participants in the user and contributor layers (Adner, 2006). Initiation risks are similar to other types of businesses and include product feasibility, supply chain access, and potential competition. Risks specific to platforms include interdependence and integration risks associated with external contributors that one has little or no control over. These risks dramatically increase as the number of interdependent intermediaries increase.

To assess these risks, Adner (2006) suggests mapping the business ecosystem to reveal insights where problems may arise. Firstly, identify all the intermediates and the innovations that must be adopted prior to the users creating value. Secondly, estimate the delays caused by integrating these intermediates and innovations into the ecosystem. These delays are often compounded by those intermediates lying farther out in the ecosystem. Thirdly, on the basis of those estimates, calculate the probability of the ecosystem materializing. This risk assessment is critical if several intermediates are depended on for developing the platform. Adner (2006) calculates the risk of four intermediates, with each having a 90% of delivering their innovation, as having a 66% chance of occurring. In this case, it is $0.9 \times 0.9 \times 0.9 \times 0.9$, which is 66%.

11.3   Strategies for an eco-commerce ecosystem

To create a thriving ecosystem, a system of trust must be built into the ecosystem layers. This trust must be constructed within the context of the business ecosystem and considered a tangible, actionable asset that is created (Covey and Merrill, 2006). The platform architecture and the curation process for data and participants must provide confidence to the users and contributors that their interactions will generate positive returns. It must also enable users to work through issues that emerge.

A common mistake that managers make is to plan out the full ecosystem with rigidly held positions, roles, and strategies of delivering the product or service to the end customer. If this approach is taken, managers tend to overlook the processes through which the ecosystem will emerge over time. A successful growth strategy accounts for the delays, challenges, and risks that are inherent in collaborative networks (Adner, 2006).

A successful platform strategy (Choudary, 2015) requires a *magnet* to bring the user groups to the platform, *a toolbox* of technology to enable the users to conduct their activities, and a *matchmaking* process to correctly connect users using valid data. It must also contain an incentive design where users experience some level of fun, fame, and/or fortune.

11.3.1   User magnet

Due to the chicken-and-egg scenario of multisided platforms, attracting initial users is challenging as values are nonexistent or relatively low prior to the engagement of two or more users' groups. In the case of the multisided shared governance (MSSG) platform, the supplier group consists of the platform managers, farmers, and land managers and the demander group consists of governments, corporation, utilities, and NGOs. The magnet for the supplier group is to account for and understand their level of sustainability with the potential that some entity seeking sustainability values will connect and purchase them. The magnet for the demanders is to procure landscape data that meets their sustainability objectives.

While both are options, it would seem that sustainability demanders associated with corporate supply chains and utilities associated with water quality trading schemes would be the most promising initial clients. The magnet in this case is affordable and readily accessible $1.0 NCU data they could acquire directly from the platform and the potential to search for an optimum location to generate credits or achieve objectives.

Government users could have many application needs depending on the agency and their objectives, but may be initially reluctant to use a nongovernment platform. Green bond financers may use the platform to assess areas of interest with low natural capital functions with the potential to increase natural capital functions. Academia may use the platform for research related to landscape management, index development and use, economic valuation schemes, and transdisciplinary research. NGOs could use the platform for some or all of the uses mentioned to meet their organizational objectives. In all cases, it becomes more attractive to users as the number of engaged sectors and organizations increases.

11.3.2 Technology toolbox

The platform architecture contains the technological infrastructure for interaction and enables users to retrieve or generate new values. The MSSG platform is based on a GIS-gridded landscape containing natural capital cells (NCCs). Each NCC is identified by its location and contains landscape data with an accompanying cache of indices to generate NCUs. This architecture becomes a basis for a smart assessment whose value can cascade down the value stream to be used for planning, implementation, portfolio development, assurance, auditing, and valuation.

The MSSG platform enables a *participatory approach*, a vital component of an assessment due to its ability to bring new information to the scientific, government, or economic community. This level of landscape intelligence is seldom attainable by any other means.

This information is then packaged as NCUs with multiscale and multiscope capabilities, allowing numerous stakeholders to interact at the scale and within their parameters of choice—such as soil quality at the field scale or carbon at a regional scale. The resulting data can be readily sorted, compiled, and queried to provide specific resource information needs to varying government and nonprofit organizations, and the private sector. Therefore, the management data itself becomes a value above and beyond the value of resource-management outcomes. This occurs because the resource-assessment data generates the landscape intelligence that can become the basis for supply-and-demand dynamics of ecoservice markets. Properly packaged, it becomes commoditized data capable of generating income from a variety of transaction types. Due to the range of transaction values and the nearly unlimited number of potential transactions, an internal platform cryptocurrency is proposed.

11.3.2.1 Cryptocurrency

A MSSG platform cryptocurrency would be an internal currency designated for commerce within the eco-commerce ecosystem. A cryptocurrency is a medium of exchange designed for the purpose of exchanging digital information through a process made possible by certain principles of cryptography (Box 11.1). Cryptography is used to secure the transactions. In most cases, it is also used to control the production of new currency units or coins as it relates to fiat currencies (Godsiff, 2015).

BOX 11.1 CRYPTOCURRENCIES

A cryptocurrency is a medium of exchange such as the U.S. dollar. Bitcoin, the first cryptocurrency, appeared in January 2009. There are more than 672 in use in 2016.

Like the U.S. dollar, a cryptocurrency has no intrinsic value; it is not redeemable for another commodity, such as gold. Unlike the U.S. dollar, however, cryptocurrency has no physical form, is not legal tender, and is not currently backed by any government or legal entity. In addition, its supply is not determined by a central bank and the network is completely decentralized, with all transactions performed by the users of the system. The term cryptocurrency is used because the technology is based on public-key cryptography, meaning that the communication is secure from third parties. This is a well-known technology used in both payments and communication systems.

Source: Murphy, E.M. et al., Bitcoin: Questions, Answers, and Analysis of Legal Issues, Washington, DC, Congressional Research Service, 2015.

Fiat currencies are those that have no intrinsic value such as being backed by a physical commodity. The value of fiat money is derived from the relationship between supply and demand rather than the value of the material it is made from.

Rather than a fiat system, the value of a MSSG platform currency would be associated with NCUs in some manner. Since there is a limited number of NCCs delineated on the surface of the earth (~6.4 trillion), the total number of cryptocurrency units could be associated with this finite and apparent number, creating a nonfiat currency based on natural capital and its capacity to produce ecoservices and ecological goods.

Some of the advantages of cryptocurrencies are low transaction costs, borderless transferability and convertibility, and trustless ownership and exchange (Tether, 2015). This allows peer-to-peer transfer of value over the Internet much like paper currency is exchanged between people, except every transaction is validated with cryptography. Cryptography protocol could simplify complex asset transfers and be able to publicly identify who currently owns a unit of property such as an NCU, and could include a record of both past ownership and other history. Such efficiencies and data reduce transaction friction by allowing individuals to directly transfer property without the use of a broker, lawyer, or notary to sign. Traditional contracts could be replaced by code that self-executes when a triggering event occurs. These characteristics provide NCU transactions at a scope and scale that is simply not possible with traditional means of transactions. A carbon impact factor (CIF) is an example of a family of cryptographic financial instruments, serialized and blockchain-enabled, that is proposed to be used to quantify and value the carbon efficiency associated with the production of agriculture commodities (Madden et al., 2015).

11.3.3 Matchmaking, incentives, and trust

If users are attracted to the platform to seek value for their ecoservices or achieve a sustainability objective, then two agreeing parties must be connected to fulfill this transaction. This matchmaking process can be accomplished through queries on what type of sustainability values are being produced or which values are needed to address sustainability objectives. The platform could identify specific areas of interest such as a watershed contributing to harmful algae blooms in the Great Lakes, a region supplying drinking water to the Des Moines Water Works service area or fields that contribute a commodity for a specific product.

Connecting corporate suppliers with land managers in a cost-effective manner is often enough incentive to encourage interaction. It is through a well-designed curation process and ease of interaction that eventually builds platform use and trust.

11.4 Ecosystem emergence considerations

An emerging business ecosystem is analogous to the process of ecological succession in biology. Grasses and shrubs create or stabilize a new space and are followed by conifers and nitrogen-fixing trees such as Alders. These, in turn, produce more fertile conditions for hardwoods to progress toward the climax structure of a rich, diverse, and stable ecosystem with substantial biomass (Moore, 2006). Ecologically, this scenario is described as a state and transition model, where each state incurs some level of stability until certain thresholds are reached and the ecosystem transitions to a new state (Stringham et al., 2003).

Thresholds may be reached through gradual succession as described by Moore or by a more disruptive environmental change, such as the introduction of invasive species, pollution, disease, fire, or flood. A threshold is a point in the transition that results in a change in the state. The new state may be a degraded, less productive state, or a highly productive, more diverse climax state. As a complex system, no one can control the process of transition or the endpoint, but one can attempt to initiate and support this shift.

Thomas and Autio (2013) proposed three phases of business platform ecosystem emergence: initiation, momentum, and optimization.

The initiation phase consists of the initial idea, resource gathering, and early operation. Early operation consists of prototypes, restricted access, sense- and rulemaking, low levels of promotion, some press coverage, and insignificant competitor activity.

The momentum phase is when the ecosystem begins to grow rapidly, driven by incoming investment, increasing numbers of participants, emerging positive network effects, aggressive marketing, industry, and societal interest, as well as competitor activity. During this phase, ecosystem growth is the main goal, with much interaction and commerce between the hub firm, users, and contributors.

The optimization phase is when the focus of activity moves from expansion to control and value appropriation. With the ecosystem established as the undisputed leader, focus now shifts closer to control of the activities of ecosystem participants and value appropriation. The rate at which the ecosystem moves through these phases is related to how well resource, technological, and institutional activities are executed within the context of the existing environment at large.

Resource activities are associated with developing organizational capacity. Technological activities are associated with platform performance and institutional activities are those efforts that establish norms of behavior, governance, and procedures within the ecosystem.

Overall, the institutional activities shape a shared understanding resulting in a collective identity of the ecosystem and assist in influencing contextual activities relating to the efforts of organizations outside the ecosystem, but within the overall environment.

chapter twelve

Enabling an eco-commerce ecosystem

Business ecosystems are pretty simple in concept (Moore, 2013). The goal is to get a lot of people to bring their creativity together and accomplish something more important than they can do on their own. Business ecosystems may be disruptive to established systems as they reduce barriers and create new avenues of entry for new players and encourage new people with new ideas, money, tools, and technologies to participate and create. In this context, the objective of enabling an eco-commerce ecosystem is to bring people and organizations together to transform the agriculture system as it *is* into the system as it *ought to be*.

12.1 Three phases of transformation

Olsson et al. (2006) identified three phases in the transformations of social–ecological systems. The first phase is *preparing the system* for the potential changes. The second phase is *navigating the transition* from the system as it *is* to the system as it *ought to be*. These first and second phases are linked by a *window of opportunity* that enables change. The third phase is about *building resiliency* in the new system.

A fictitious scenario, based on the objectives of the Electric Power Research Institute, Inc. (EPRI) water quality trading case study, will be used to examine these three phases of transformation in the process of emergence of an eco-commerce ecosystem.

12.1.1 Preparing the system

Preparing the system for transformation consists of creating new networks, acquiring new knowledge, and identifying new values (Olsson et al., 2006). Preparation can occur organically as organizations seek solutions and develop new relationships that may lead them to adaptive comanagement processes. Adaptive comanagement occurs among network collaborators and is one precursor that enables adaptive, shared governance (Plummer et al., 2013).

In addition to this social component, a more deliberate or controllable preparation consists of creating the space for engagement. In the case of the eco-commerce ecosystem, preparation requires the development of a multisided platform. This engagement space supports the adaptive comanagement processes and enables new ways to interact and conduct transactions. The development of a multisided shared governance (MSSG) platform is considered a tame problem that can be engineered, tested, and refined, while the social, adaptive comanagement process is a wicked issue that must emerge from the community.

12.1.1.1 *Adaptive comanagement*
Adaptive comanagement is applied to achieve outcomes which are not possible separately. It combines the ecological and working knowledge of farmers and scientists in mutual learning systems, drawing on both farmers' experience and scientists' knowledge (Plummer et al., 2013).

Folke et al. (2005) described adaptive comanagement as a means to adaptive governance. It relies on the collaboration of a diverse set of stakeholders, operating at different levels, often through networks from local users to regional and national organizations. It is a sharing of management power and responsibility among user groups or communities, government agencies, and nongovernmental organizations.

These adaptive comanagement characteristics were observed in the majority of the case studies. As Figures 8.9 and 8.10 show, regardless of the governance styles used or the source of the primary governance actors, the overall delivery of the projects was quite inclusive as shown by the more "rounded" governance compasses. Of the 11 case studies, three (The Sustainability Consortium [TSC], Field to Market [FtoM], and United Suppliers, Inc. [USI]) are converging on a more formal adaptive comanagement strategy of sorts as they develop closer relationships to gather data and acquire the new knowledge needed to account for and achieve sustainability.

These adaptive comanagement activities prime the opportunity of transformation toward adaptive governance systems, but do not guarantee it (Olsson et al., 2006). Chaffin et al. (2014) noted that although it is essential that a diverse array of vested stakeholders participate, individual leadership and trust building among stakeholders at the local level are what drive the emergence of adaptive governance. To build this trust and prepare potential leaders for the emergence of an eco-commerce ecosystem, interaction, value, and transactions must be realized at the local level.

12.1.1.2 *Minimal viable platform*
Enabling an eco-commerce ecosystem requires the MSSG platform. An ecosystem emerges from a platform and a platform grows from a single interaction. To initiate an ecosystem, Choudary et al. (2015) advise to first

create a minimal viable platform to demonstrate its capacity to support one interaction. Eventually, user groups must interact with each other to create new values, but initially it is advisable to connect one user that can obtain value from the platform.

The first step to developing a minimal viable platform begins with identifying how a new type of interaction may address a problem in need of resolution. For example, OpenTable is a platform to connect potential diners with restaurants. For the first interaction, OpenTable developed software to assist individual restaurants to seat and track diners. Individual restaurants were able to increase their efficiencies, but were not able to interact with potential customers. As more restaurants adopted the software, a critical mass of users *prepared* the system to transform from an isolated software program to a multisided platform.

To prepare the agriculture system for a transformation, a minimal viable platform is needed to allow one or several stakeholders to account for and identify methods to improve agriculture landscape sustainability. Such a minimal viable platform would contain a *LandIT* layer and a *LandApp* layer. A LandIT (information technology) layer contains the geographic information system (GIS)-based landscape data, a landscape delineated with natural capital cells (NCCs), and a depository of curated indices to calculate the *supply* or the *capacity-to-supply* specific natural capital units (NCUs). The LandApp (application) layer enables the user to design and analyze various landscape management scenarios to determine their effects on S1.0, S1.5, and S2.0 values. These data and computational capacities are readily available through private and public entities.

12.1.1.3 *A first platform interaction*

The minimal viable platform enables the first platform interaction. The first interaction with the MSSG platform consists of delineating an area of interest on the NCC-gridded landscape and then identifying the existing S1.0 supply (baseline) and creating an S1.0 portfolio. This is followed by identifying potential areas and activities to create S2.0 credits, and an estimate of the costs. *The Utility Company* (TUC), a fictitious entity based on the EPRI case study scenario, will be used to illustrate the interactions.

One complete interaction includes the following objectives and tasks:

1. Identifying sustainability supply
   a. Delineate an area of interest (group of NCCs)
   b. Gain access to NCC landscape and management data
   c. Select indices from platform depository
   d. Apply indices to NCCs to generate NCUs
   e. Generate a S1.0 portfolio (ecoservice output)

2. Enhancing the portfolio
   a. Identify areas for potential ecoservice improvements
   b. Calculate costs and benefits of applying landscape management practices and structures
   c. Create S2.0 portfolio options
3. Propose transaction
   a. Submit S1.0/2.0 offer to land owner
   b. Begin transaction/propose payment
4. Assurance
   a. Monitor NCUs on a periodical basis (seasonally, yearly)
   b. Administration related to certification, payments, assurance, certainty, etc.

12.1.1.3.1 Identifying a new sustainability supply An interaction begins by TUC delineating an area of interest on the platform map to determine the ecoservice supply (#1a–e). The *LandIT* layer is applied to the delineated area, providing access to a cache of indices. Indices are selected and applied to the NCCs to generate NCUs calculated on a 0–100 scale. These NCUs represent an ecoservice supply of the delineated area based on *superficial* landscape data such as topography, soil, vegetation cover, and climate (Figure 12.1). This superficial landscape data is available in variable quantities and periods of time, but is becoming more available via satellite and other data sources in many regions.

With this first platform interaction TUC acquires S1.0 NCUs: landscape intelligence on the supply of ecoservices based on superficial

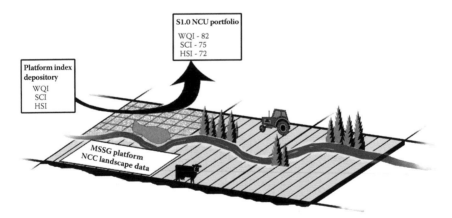

Figure 12.1 The first interaction with the platform is by The Utility Company (TUC) seeking S1.0 values associated with superficial landscape data. TUC delineates an area to generate NCCs and applies specific indices from the depository to calculate NCUs. These NCU values are compiled to create an NCU portfolio.

landscape data. To procure this data, a transaction occurs between TUC and the MSSG platform managers. This data is readily available on the open market, and so the value for TUC to engage with the MSSG platform is not necessarily just a convenient data package, but the opportunity to engage with other platform users to create new values. The ultimate cost of NCC and NCU data is based on typical market negotiation of supply and demand data, but it would presumably be quite low. To ensure transaction costs can be kept at a minimum, an internal platform cryptocurrency would facilitate transactions, track ownership, and apply other traits of electronic currencies to address legal and accounting requirements.

12.1.1.3.2 Enhancing the portfolio With S1.0 landscape intelligence and LandApp, TUC can identify where changes in landscape management could produce S2.0 NCU values. Ideally, TCU would procure S1.5 data that consists of detailed management practices of the land manager, but since this is the first interaction no other user groups are participating. Enhancing the portfolio (#2a–c) begins with LandApp layer algorithms applied to the LandIT layer to identify areas that would respond positively to conservation management options. In this layer, the user can create conservation practice designs and estimate the costs to generate S2.0 values such as capturing and treating a volume of water. In Figure 12.2, an area is delineated as a proposed location to store and treat agriculture runoff. The potential water quality credits may be calculated using an

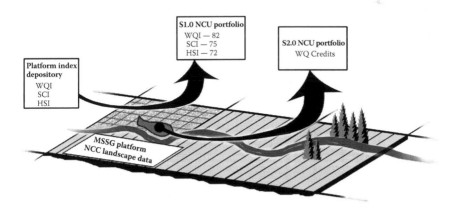

Figure 12.2 The next step of the first interaction is applying the LandApp layer to determine areas to best generate S2.0 values. In this example, an area is delineated next to the stream to store and treat agriculture runoff. The LandApp layer is able to provide options based on basin size, treatment potential, and the rate of release to calculate the number of potential credits and the costs of constructing these plan options.

index related to the number of cubic meters treated. In some cases, direct measurement and analysis is cost-effective for S2.0 credits since the outputs from the treatment area are no longer diffuse, and discharges are controllable and measurable at a weir or piped outlet.

12.1.1.3.3 Propose a transaction   The first interaction consists of only one user with the platform, and so, there is not the potential of a transaction with another user group. Similar to the OpenTable process, TUC finds value interacting with the MSSG platform as a data and planning resource. To generate new values beyond this single group interaction the platform must have a process to identify values for a potential transaction. In this example, TUC seeks water quality credits for a water trading program. A transaction proposal may consist of an offer to acquire S1.5 data for a particular area identified as ideal to generate S2.0 values (#3a–b).

12.1.1.3.4 NCU assurance   Assuring ecoservice production and the legitimacy of NCU transactions may occur by remote sensing, accessing GIS databases, and viewing management records, or by on-the-ground confirmation. Since trust and legitimacy are critical components of business ecosystem creation and facilitation (Thomas and Autio, 2013), curation of users, contributors, and data is an important aspect of assurance. This first MSSG platform interaction produced new knowledge and provided a means to account for sustainability values, but one-sided interactions have limited value and will not cause an ecosystem to emerge.

12.1.1.4   The next platform interaction

A platform has the capacity to support an ecosystem if the next platform interaction is by a user group with the potential to conduct a transaction with the first user group. New and enhanced values can emerge from a platform when the first interaction is followed by the next interaction connecting users and contributors (Choudary et al., 2015). To entice platform interactions leading to transactions, value must be realized by both parties. Some multisided platforms shift the costs of the transaction to one user group to make the entry costs for another user group affordable or free. The OpenTable example uses this strategy to provide the service for free to potential diners. Restaurants carry the costs of both groups as they are motivated by potential customers.

In the context of the MSSG platform, the land manager user group adds value by providing additional management data and assurance certain land management strategies are applied and maintained. Nutrient management, tillage, and other cultural practices, only available through practitioner knowledge, are incorporated into the indices to create more precise S1.5 portfolio information. This additional value is illustrated in Figure 12.3.

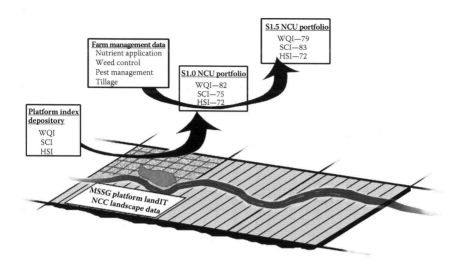

Figure 12.3 A second platform interaction by a land manager incorporates new information to the S1.0 NCU portfolio to create an S1.5 value. The S1.5 data is more precise as it incorporates management data into the superficial S1.0 landscape data. This interaction begins the matchmaking process needed to transact data and values and to provide assurance sustainability objectives will be met and maintained.

The motivation for the land manager to engage with the platform is to identify their level of landscape sustainability relative to emerging sustainability demands. Similar to the OpenTable format, the MSSG platform could allow free or very affordable access for land managers to generate S1.5 portfolios. The land managers are the user group that provides a primary value to the platform by incorporating new data for a more refined NCU calculation and are the only user group that could provide assurance that landscape management strategies will be applied and maintained. Without land management data and an assurance process, the platform is far less useful and less valuable and legitimacy and trust of the data is suspect. The land managers' user group increases the value of the platform by (1) incorporating refined management data for more precise LandIT calculation (S1.5 NCUs), (2) vouching for S1.5 values, and (3) being a critical partner for S2.0 investment opportunities for S2.0 values.

Using this strategy, TUC (first interaction), seeking S1.0 and S1.5 values and potential S2.0 values, could pay the transaction costs while the land manager, presumably conducting the next interaction, would have free or discounted access.

12.1.1.5 A prepared system

Preparing a social–ecological system for change is often a long process consisting of technical, scientific, political, and social changes. The social changes can take decades. The sustainability movement of the 1980s ushered in new stakeholders for all aspects of agriculture production (Sampson et al., 2013). The new knowledge and support eventually led the United States Department of Agriculture (USDA) to change the Soil Conservation Service (SCS) to the Natural Resources Conservation Service (NRCS) with a broader mission in 1994. Corporations, industry, utilities, and nongovernmental organizations (NGOs) have initiated similar changes as represented by the case studies. They emerged from the social and political environmental movement of the previous decades.

To leverage this cooperative social capital, a MSSG platform was proposed to generate NCCs, NCUs, and ecoservice portfolios to create a new supply and a process to combine the low and diffuse ecoservice values and reduce and share the high and multiple transaction costs. It is through transactions that a transformation from the system as it *is* to the system as it *ought to be* will occur. With this *transaction* prepared system, sustainability leaders can initiate an eco-commerce ecosystem when a *window of opportunity* opens.

12.1.2   Window of opportunity

Once new system dynamics are prepared, a shift in governance toward adaptive governance requires a window of opportunity (Olsson et al., 2006). Such windows may appear as a significant boost in capital or legitimacy in a current effort, a shift in policy, a disruptive political election, a natural event such as an ecological disaster, or the transformation of a previously informal network into a formal governance organization (Chaffin et al., 2014).

Olsson et al. (2006) describe a window of opportunity as when problems, solutions, and politics come together at a critical time. They cite an Everglade case study where an algae bloom in Lake Okeechobee triggered a transformation that was preceded by networks and leadership associated with seeking new solutions. The extent to which the transformation of entire governance regimes is possible appears to be related to the scale at which the crisis most clearly manifests itself and how it is perceived in relation to the scope of change possible. In the case studies analyzed, windows of opportunity appeared and inspired leadership to address a perceived crisis and transform governance in some manner.

12.1.2.1   Case study windows

In 2001, the Minnesota Milk Producers Association recognized the increase in regulatory scrutiny and new sustainable supply chain efforts as an opportunity to pursue a change in how dairy farmers received technical assistance and how they accounted for farm sustainability. They approached the Minnesota Legislative Citizens Commission on Minnesota Resources to fund their environmental quality assurance program (MMEQA) to resolve these issues.

Total resolution of the issue did not occur, but going through this first window of opportunity created additional opportunities. As the MMEQA concluded in 2009, the Minnesota Legislature supported the Livestock Environmental Quality Assurance (LEQA) program in 2010–2011 which subsequently led to two separate projects: the Agriculture Water Quality Certainty Program (AWQCP) in 2012 and the Agriculture Environmental Quality Assurance (AgEQA) in 2013. The successes of the previous projects revealed potential solutions and created additional political momentum to open more windows.

The windows of opportunities for the national-scale case studies are less defined but may have been opened from a compilation of issues associated with water, climate, and the environment. These perceived crises were accompanied by growing social awareness, government regulatory responses, and corporate sustainability efforts. These regional and global crises are often accentuated by more local crises such as harmful algae blooms (HABs) creating drinking water concerns in the Great Lakes region and agriculture runoff resulting in high cost of water treatment such as that noted in the Des Moines Water Works watershed. The potential for windows of opportunities seem unlimited.

12.1.2.2   Too many "little" windows?

In fact, these emerging crises at the various glocal scales have the effect of creating many windows of opportunities for people and organizations to convene, evolve, and adapt. Organizations are compelled to address their issues within the context of their resources and interests. As they seek solutions they likely follow one of the several governance pathways, create new accounting and valuation systems, and adopt governance styles that may or may not conflict with organizations with similar objectives.

These are the consequences of many organizations responding to glocal sustainability issues without access to a common platform or engagement within a solution ecosystem (Eggers and Muoio, 2015). The wicked challenge with agricultural landscape sustainability is to create one window of opportunity that supports the convergence of these resources, or to enable these many windows of opportunity to converge to a common platform and a single ecosystem.

A convergence of sorts is occurring with the three case studies of the TSC, FtoM, and USI projects. TSC and FtoM have signed a memorandum to harmonize metrics, encourage data platform interoperability, and collaborate together on innovation projects. This partnership will enable Field to Market's metrics to be used in reporting against TSC's key performance indicators for several commodity crops (Field to Market, 2014b).

USI does not have formal agreements with TSC and FtoM but is working with the Environmental Defense Fund to aggregate metrics that address the sustainability objectives of Walmart, a major TSC member.

While this convergence is an important trend, other sectors such as government, utilities, insurers, and various NGOs remain committed to their unique efforts and their own windows of opportunities.

12.1.2.3 Too big of a window?

It is difficult, if not troubling, to imagine a global event with the potential to create the social awareness and political will for a new, inclusive system to coalesce. Perhaps it would take a so-called black swan event (Taleb, 2007): those unlikely, unanticipated, and relatively large-scale events that cause swift changes in perspectives, or a black elephant event. London-based investor and environmentalist Adam Sweidan states a black elephant is a cross between "a black swan" and the "elephant in the room." It is a problem that is visible to everyone, yet still no one wants to address it even though it will have vast, black swan-like consequences. Sweiden lists environmental black elephants gathering out there such as global warming, deforestation, ocean acidification, mass extinction, and massive fresh water pollution (Friedman, 2014).

Unfortunately, as it relates to water, climate change, and other sustainability issues, a window of opportunity big enough to cause a mass social and political convergence may signal a tipping point for natural systems that are unable to return to the desired state.

12.1.2.4 Opening a WQ trading window

A window of opportunity opens where political interest, known problems, and possible solutions come together at a critical time (Olsson et al., 2006). Water quality trading is one such promising and plausible opportunity. Water quality trading markets in the United States have their origin in the Clean Water Act (CWA) of 1972 with the basic objective of restoring and maintaining the integrity of the nation's waters and establishing total maximum daily load (TMDL) regulations to limit pollutants in impaired waterways.

Despite the uncertainty associated with water quality improvements and the lack of scalable success in water quality markets, political interest continues to grow. This growing interest is, in part, due to the high costs, political resistance, and ineffectiveness associated with regulating nonpoint source pollution from agricultural landscapes.

12.1.2.4.1 *Political interest* A September 2015 National Workshop on Water Quality Markets in Lincoln, Nebraska was cosponsored by the USDA and EPA to figure out how to enable water quality trading markets to overcome the ongoing challenges of creating viable markets. EPA officials support water quality trading as an option. Ellen Gilinsky, EPA senior advisor on water quality, stated regulations alone will not achieve EPA's goal of a 45% nutrient reduction for the Gulf of Mexico and so we are going to need creative solutions (Clayton, 2015).

Prior to the workshop, the USDA announced $2 million in Conservation Innovation Grants to support water quality trading markets, which included among others a $400,000 grant to the Great Lakes Commission to create a water quality trading framework to address HABs, a $700,000 grant to the Iowa League of Cities to develop a framework for water quality trading in Iowa to support the State's Nutrient Reduction Strategy, and a $300,000 grant to EPRI to develop and execute trades of combined eco-services for water quality and greenhouse gas emissions reduction credits (USDA, 2015).

12.1.2.4.2 Persistent market problems Water quality trading has experienced persistent problems. In the March 2009 Water Quality Trading Programs report, the World Resources Institute identified 57 water quality trading programs worldwide with 51 of those residing in the United States. Less than a fourth of them finalized their trading program design with fewer having active trades (Selman et al., 2009). In addition, the few water quality trades that are transacted are not of the kind considered with a commodity-like market. Most are bilateral agreements that were based on ad hoc criteria and direct negotiations. These trades do not provide many insights to help guide the development of a trading scheme needed to significantly reduce the cost of achieving water quality goals in most parts of the country (King and Kuch, 2003).

Typical of most water quality trading markets, EPRI also required government and private sector subsidies in its development stage (EPRI, 2012). In May 2015, an EPRI water quality trading auction was cancelled after an initial April auction was postponed and the required entry level bids were lowered from $20,000 to $5000 (Fox, 2016). An official report identifying the reasons for the postponement has not yet been produced, but presumably the uncertainties associated with the trades and high transaction costs were a contributor.

12.1.2.4.2.1 Uncertainty in WQ trades Uncertainty in water quality trading is related to the difficulty in measuring water quality benefits of practices, the accountability of the trade, and the market's effect on improving the overall watershed quality. Each of these remains a critical obstacle to developing markets. Beyond the uncertainty of the effect of the practice, uncertainty arises on whether practices are installed or maintained. Food and Water Watch raised this issue with a power plant in Pennsylvania that bought 80,000 pounds of nitrogen credits, but questioned if there was a method to determine whether farms actually reduced 80,000 pounds of nitrogen applications (Neeley, 2015).

A third uncertainty is related to the overall effect the water quality trades have on the watershed condition—even if the first two uncertainties are addressed. Since all portions of the landscape contribute to the water quality, but only a very small percentage is accounted for in trades, uncertainty arises as to whether the remaining landscape is managed in a manner that does not eliminate the benefits generated by the water quality trades. If the trading program and the buyer's motivation is to address watershed improvement objectives, but the overall water quality does not improve, it is uncertain whether the discharger would eventually have to upgrade its facilities in the future.

12.1.2.4.2.2   High transaction costs   Transaction costs for water quality trading markets are typically high. Ex-ante or market setup costs may span several years and include concept review and approval, baseline assessments, setting objectives, negotiating agreements, public outreach, and communication (Cherry et al., 2007).

Costs of market operation include information gathering, negotiation, contract formulation, monitoring, and enforcement. Strategic costs are associated with addressing noncompliance, free-riding, and unknowns associated with new ventures. After-market or ex-post costs are associated with monitoring performance, sanctioning and governance, and renegotiation when the original contract is unsatisfactory. Transaction costs are not only financial. Time and other in-kind contributions should be measured and, wherever possible, monetary values of these inputs should be calculated.

The market setup and delivery costs incurred by the EPRI (2012) effort included the following process:

1. EPRI enters into agreements with state agencies.
2. State agencies enter into agreements with SWCDs.
3. SWCDs conduct outreach with landowners to secure their participation.
4. SWCDs review BMP projects for eligibility and value, and make recommendations to EPRI.
5. EPRI selects and approves BMP projects to receive funding.
6. SWCDs enter into agreements with selected landowners.
7. Landowners implement BMPs with technical support and oversight from SWCDs.
8. State agencies serve as verifiers to monitor, inspect, and verify BMPs.
9. SWCDs register BMPs and associated credits using a credit registration and tracking system.
10. Verifiers conduct annual monitoring, inspection, and verification of BMPs.

To transact a water quality credit, the EPRI followed these steps:

1. Register Point of Generation credits using credit registration and tracking system.

2. Set aside a credit pool for reserve/assurance.
3. Post remaining credits for sale.
4. Buyers submit purchase requests.
5. Apply trading equation to account for watershed specific nutrient saving between the point of credit generation and the point of use.
6. Credits are transacted.

The numerous steps required to address the uncertainty inherent to most water quality trading programs greatly increases the transaction costs. Such an arduous process for each entity contributes to the high and multiple transaction costs. Trades will not occur if the transaction costs exceed the benefits of a potential trade (Abdalla et al., 2007; Brown et al., 2007).

12.1.2.5 Entering a WQ trading window

Leadership is needed when entering a window of opportunity and to begin transforming socioecological governance. The combination of the emergence of the political support for water quality trading and the current challenges associated with water quality trading creates a dynamic tension and the energy needed to "shoot the rapids" and to navigate the transition (Olsson et al., 2006) toward an eco-commerce ecosystem.

12.1.3 Navigating the transition

With the system prepared and the potential of a window of opportunity, a transition toward adaptive governance and a functioning eco-commerce ecosystem can be initiated. Olsson et al. (2006) uses the phrase "shooting the rapids" as an organizing metaphor because it is analogous to the changes and turbulence observed in the period of transition. These *rapids* are fueled by the social momentum built during the system preparation phase and released by the opening of a window of opportunity such as to develop a water quality trading scheme and a new governance model.

Folke et al. (2005) observed social capital is built up in emerging formal and informal networks creating new platforms, arenas, or space for collaboration. These new networks facilitate information flows, identify knowledge gaps, and create new nodes of expertise (Olsson et al., 2006). Shadow networks (informal groups of participants), emerge where formal networks and many planning processes fail (Gunderson, 1999). These fundamental changes in structure, function, and relationships within socioecological systems lead to new patterns of interactions among actors, and new dynamics and outcomes (Patterson et al., 2015).

A "shooting the *eco-commerce ecosystem* rapids" scenario begins by a participant interacting with the MSSG platform to initiate water quality trade. The goal is to generate transactions with S1.0, S1.5, and S2.0 values. Since platform managers cannot simply command stakeholders to participate, the first step toward these transactions has to employ informal governance mechanisms such as trust, legitimacy, and leadership to encourage interaction (Thomas and Autio, 2013).

12.1.3.1 *Navigating the first WQ transaction*

A first transaction would use the process outlined for the minimal viable platform interaction and involve the platform managers, TUC, and a land manager (E.E. Farms). In this fictitious scenario, TUC is motivated to address their water quality discharge requirements through water quality trading, similar to the participants in the EPRI case. Water quality trades would consist of credits generated by reducing nonpoint source pollution (runoff from agricultural lands and other diffuse sources) from E.E. Farms. These credits would be used to mitigate the point source discharges of TUC.

To generate this credit, the first water quality transaction component would consist of an S1.0 trade between the MSSG platform and TUC (#1–3) and an assessment for viable S2.0 credits (#4). The second transaction component would consist of an S1.5 trade between TUC and E.E. Farms (#5–6). The third component is an S2.0 trade between TUC and E.E. Farms (#7–8). Figure 12.4 illustrates the flow of data and money with each step outlined below.

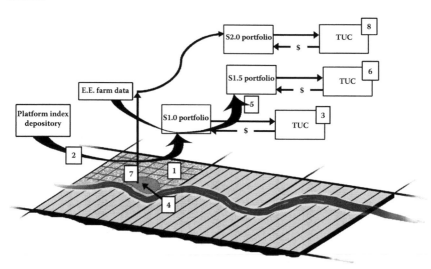

Figure 12.4 Navigating the first water quality trade on the MSSG platform involves four phases (and eight steps). The Utility Company (TUC) first acquires S1.0 data from the platform (#1–3) and secondly assesses the potential to create S2.0 credits (#4). With this information, TUC procures S1.5 data from E.E. Farms (#5–6). With assurance of farm management, TUC proposes an offer to generate S2.0 credits (#7–8).

1. TUC interacts with the MSSG platform to delineate NCCs representing the area of interest.
2. TUC selects indices from the platform depository and applies them to the LandIT layer to generate NCUs.
3. TUC purchases the NCC and NCU data from MSSG platform managers as a S1.0 transaction.
4. TUC queries the LandIT layer for specific areas of interest to generate S2.0 credits.
5. TUC initiates contact with E.E. Farms to seek S1.5 data.
6. TUC purchases S1.5 data from E.E. Farms to establish more refined NCU data.
7. TUC applies LandApp layer to generate S2.0 data, identify costs to implement water storage and treatment and S2.0 credits.
8. TUC offers contract to E.E. Farms to provide S2.0 credits.

This process is very similar in content to the EPRI transaction process, but since it is conducted within the MSSG platform much of the activity is electronically streamlined, accounted for, and recorded. It significantly reduces upfront costs of assessing the landscape and connecting user groups. It has far greater transaction efficiencies for S1.0 and S1.5 transactions and locating S2.0 credit opportunities. Generating S2.0 credits can be an expensive and time-consuming process of negotiation and construction, depending on the project objectives. But even this process would be streamlined with the capabilities of the LandIT and LandApp layers. The MSSG platform provides this new space as aptly described by Scharmer and Kaufer (2013) as a *cocreative eco-system* to enable cross-sector innovation and engagement in the new era of organizational governance. It is within this space that negotiation can occur to seek out values for agriculture landscape sustainability.

12.1.3.2 Transforming water quality trading

For water quality trading to become a significant force in addressing local, national, and global water resources, it is apparent that the current models must be transformed into more market-like systems. The MSSG platform transforms water quality trading through the following characteristics:

1. It contains the means to account for and monitor baseline management of the landscape contributing to the S2.0 project to reduce uncertainties. An entity seeking S2.0 water quality credits has the option to establish S1.0 or S1.5 baseline criteria at the scale (field, farm, or watershed level) they desire.
2. The water quality credits based on S2.0 values are associated with tangible outputs such as the quantity of water stored, treated, and

released. This differs from traditional and typical water quality trades that may be based on a wide variety of cultural and structural conservation practices including S1.0- and S1.5-type values associated with the ecoservice outcomes.

3. The S2.0 approach addresses the scientific concern associated with conservation practices and their effect on water quality during low and high flows. The basis for water quality trading is to allow a higher pollutant load to be discharged at the point source in exchange for treatment practices for nonpoint sources. In periods of low flow, where little rainfall or runoff occurs, the discharge has a more pronounced effect. In some low-flow cases, municipal, industry, and utility discharges are the primary source for flows (Stine, 2015). In these cases, dischargers mitigating their treatment requirements through nonpoint source practices (functioning only during runoff events) are not necessarily mitigating the stream needs during low-flow conditions. In the case of high rainfall and high-flow levels, the point source discharges are significantly diluted and the number of conservation practices relative to the size of the watershed is generally extremely low. Requiring water quality trades to consist only of S2.0 output-type trades within the context of an S1.5 management level addresses the water quality and quantity issues associated with low-flow and high-flow conditions.

4. It enables a shared governance approach to allow multiple entities to interdependently value sustainable landscape management to create ownership and enable *symbiotic transactions*.

12.1.3.3 *Enabling symbiotic transactions*

Initiating the first transaction generates appreciation for Olsson et al. (2006) "shooting the rapids" metaphor for transitioning toward an adaptive governance strategy. Preparing the system for change is essential for leadership to gain the support and confidence to "shoot the rapids" if and when a window of opportunity is opened. To move beyond this initiation phase, Thomas and Autio (2013) state that a business platform ecosystem must go through a momentum phase where the ecosystem begins to grow rapidly as more participants understand and use the ecosystem. For the eco-commerce ecosystem to gain momentum, additional sustainability stakeholders must find value by engaging with the MSSG platform and the initial user groups.

12.1.3.3.1 Symbiotic demand As described in Chapter 9, symbiotic demand is the result of multiple entities demanding sustainability values within a common forum so that the low and diffuse ecoservice values can be readily combined and the high transaction costs can be reduced and shared. The six case studies from Figure 9.4 will be used to illustrate how disparate stakeholder values and processes can be transformed into a symbiotic demand outcome. The NRCS, AWQCP, CBay, USI, TSC, and EPRI projects are efforts initiated by different governance actors using different accounting strategies and governance styles. Each project has a vested, yet separate interest in agriculture landscape sustainability and presumably, none are able to address the wicked problem of landscape sustainability on their own accord.

Figure 12.5 illustrates how these organizations have the potential to interdependently interact to acquire S1.0 and S1.5 data and achieve their sustainability objectives. The following trades generate S1.0 data and S1.5 data: club goods that are potential units of trades. This process is explained in the six steps below:

1. The MSSG platform contains S1.0 data that is available to each of the six case study participants. The EPRI procures the S1.0 data, but since the other five are in different sectors and are only seeking data

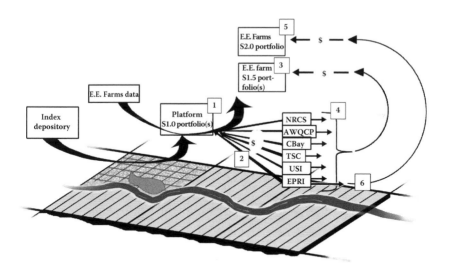

Figure 12.5 Symbiotic demand occurs when multiple demanders (NRCS, AWQCP, CBay, TSC, USI, and EPRI) are able to (#2) procure S1.0 (#1) and S1.5 (#3) values for a particular parcel of land, thereby potentially reducing the cost for each demander. Ecoservices calculated as S1.0 and S1.5 NCUs act as exclusive and nonrival club goods whose value can be shared. S2.0 NCUs (#5) act as exclusive and rival private goods and are only purchased by one entity: EPRI (#6).

representing landscape sustainability data (not credits) this data is also made available to them. Pricing schemes may be developed to encourage enhanced cooperation among multiple entities and are part of the negotiation phase to identify price.

2. Each entity procures S1.0 data as a club good that is customized with specific indices to address their sustainability needs.

3. E.E. Farms incorporates their data onto the platform along with S1.0 data to produce S1.5 portfolios to meet the specific needs of the sustainability demanders.

4. Each of the six entities procure S1.5 data from E.E. Farms via the MSSG platform as a club good.

5. E.E. Farms establishes S2.0 credits for EPRI as a private good with contract stipulations to maintain a S1.5 portfolio at a specific management baseline.

6. EPRI procures S2.0 water quality credits.

Demand for sustainability data is able to materialize into symbiotic transactions due to the nature of club goods. S1.0 and S1.5 NCUs are commodities that are exclusive and nonrival. In other words, they have ownership characteristics like private property, but can be distributed to and/or used by multiple parties without diminishing their value for any individual party.

Limiting the number of entities with access to S1.0 and S1.5 data for a particular parcel of land would be appropriate for those sourcing a sustainable-produced commodity, but other types of sectors could be unlimited in their access. For example, a specific kilogram of corn can only be sourced from one location and used by one processor. In this case, only one food processor claiming a sustainable product produced from this kilogram of corn could legitimately use this S1.0 and S1.5 data. On the other hand, governments at multiple levels have a variety of interests related to wildlife, water quantity and quality, and soil health. To accomplish their objectives or to account for their progress, each agency would need access to S1.0/S1.5 values. Such is the case for the NRCS, AWQCP, and CBay projects, where each seeks common data for specific reasons and uses. None of these S1.5 demands would need to be restricted from a club good perspective and, in fact, restricting access to S1.5 data for government projects would debilitate many government programs that need landscape data to determine needs and status. Of course access to proprietary data may be restricted for other reasons, but those would be in the interest of the suppliers. These type of issues and decisions need to be spelled out clearly to ensure that trust and legitimacy are key aspects of the platform and the sustainability portfolios. These transactions must address the concerns of double-counting, additionality, and permanency at the landscape scale (Box 12.1).

BOX 12.1 CONCERNS OF DOUBLE-COUNTING, ADDITIONALITY, AND PERMANENCY IN ECOSERVICE MARKETS

Double-counting occurs when multiple ecoservices are produced and credited due to a single land management activity (Cooley and Olander, 2011). *Additionality* relates to whether activities generate new ecoservices (Bennett, 2010; Cooley and Olander, 2011). *Permanency* relates to if the ecoservice will be produced permanently or for a defined period of time (Halldorsson et al., 2015; Hejnowicz et al., 2014). Incorporating such requirements can be an impediment to the adoption of ecoservice markets and are specifically difficult with the traditional project-by-project accounting that is cumbersome and costly (Bennett, 2010).

The MSSG platform and eco-commerce ecosystem approach accounts for the production of ecoservices in a manner aligned with the production system of agriculture commodities: on a seasonal or yearly basis. This approach allows ecoservice markets to apply permanency, double-counting, and additionality on the same relatively level playing field as commodity markets that trade provisional eco-goods such as wheat or corn.

For example, if E.E. Farms (Figure 12.5) has produced wheat for each year during the last century, the current year's production of wheat is not limited by concerns of double-counting if the farmer sells the accompanying straw, or by additionality issues because the farmer has grown it for the previous 99 years, or by permanency, that the wheat will not last forever or a specific length of time. The commodity of wheat and its value is recognized as a flow of eco-goods from the landscape whose value of production is dependent on that flow on a seasonal or yearly timeframe. It is this framework that the MSSG platform accounts for in NCU production. NCUs are identified as specific ecoservices flowing from a specific NCC during a specified time frame. In this manner, they become part of the economic and ecological system as they are generated. Accounting for the NCUs on the MSSG platform provides buyers and sellers with relatively real-time production rather than a disparate project-by-project approach.

12.1.3.4 Glocal sustainability portfolios

The S1.0 and S1.5 NCUs are the shared values of agriculture landscape sustainability. They exist simultaneously in portfolios of various stakeholders at local, regional, national, and global scales. S1.0 and S1.5 values may exist simultaneously in a farmer's (supplier) portfolio as well as in the portfolios of several sustainability demanders. The S1.0 and S1.5 NCUs are generated during specific periods of time and so the portfolios have both spatial and temporal value. As the MSSG platform becomes more sophisticated, NCUs can be readily tied directly to climate and weather events and their value would fluctuate accordingly. They are essentially natural capital commodities whose production is continuous and influenced by a wide variety of factors.

12.1.3.5   Antithesis of the tragedy of the commons

If the MSSG platform is able to provide a cost-effective means for organizations to account for and address their sustainability demands, the deeper economic question is if the MSSG platform and eco-commerce ecosystem is robust enough to address the "tragedy of the commons,": the open access issue made famous by Hardin (1968) and a source of the wicked problem associated with agricultural landscape sustainability.

In review, the "tragedy of the commons" denotes a situation where individuals acting independently and rationally according to their self-interest behave contrary to the best interests of the whole group by depleting some common resource. Applied to today's agricultural system as it *is*, the "tragedy of the commons" could be described as "the negative economic and ecological consequences of an individual's land management activity, whose is able to avoid a significant portion of the costs of that activity and pass the remaining costs onto other individuals and entities."

Within that context, the opposite of the "tragedy of the commons" could be defined as "the positive economic and ecological consequences of an individual's land management activity, who is compensated for a significant portion of the benefits of that activity by multiple individuals and entities." This is the scenario depicted in Figure 12.5 where value flows from the sectors benefiting from natural capital improvement to those that manage the landscape. Free-ridership is reduced or eliminated as those without the data cannot legitimately claim sustainability. Consumers and citizens do not directly engage in eco-commerce at this level, but presumably any costs associated with commercial products and services and governmental programs would be passed along to the consumers and taxpayers.

This economic solution for the "tragedy of the commons" is now possible due to the capacities of multisided platforms and the emergence of e-commerce ecosystems. Alibaba's business strategy of creating and supporting the structure, functions, and processes of an e-commerce ecosystem is applicable to enabling eco-commerce transactions. It is through harnessing this complexity that an eco-commerce ecosystem is enabled to support the transactions and compensate the land managers creating the structure, functions, and processes of the agroecological ecosystems.

12.1.4 Building resiliency

Beyond initiating and navigating this transition toward an adaptive governance model, Olsson et al. (2006) third phase of adaptive governance is building resiliency. Resiliency is the potential of a system to maintain its structure and function in the face of disturbance, and the ability of the system to reorganize following disturbance-driven change (Carpenter et al., 2001).

An agricultural system that compensates land managers for producing ecoservices would remain resilient if motivation for producing ecoservices were embedded in the economics and if adaptive comanagement were used to direct local actions. These two microeconomic stabilizing forces must be accompanied by the ability to navigate the larger sociopolitical and economic environment (Olsson et al., 2006).

The MSSG platform and eco-commerce ecosystem is based on motivating farmers and land managers by creating an ecoservice valuation structure. It supports organizations trending toward more inclusive governance styles and adaptive comanagement strategies. To build resiliency, the eco-commerce ecosystem must engage in and with the glocal sociopolitical and economic environment. To optimize an emerging ecosystem, Thomas and Autio (2013) state the platform firm must be in a powerful position, maintain strong bargaining power in relation to other participants in the ecosystem, and provide a compelling vision for the future.

12.1.4.1 Compelling visions

In a socially integrated world, a compelling vision that includes an incentive design where users experience some level of fun, fame, and/or fortune is necessary for a successful platform (Choudary et al., 2015). This concept is aligned with successful sustainability science where it is used to understand how to manage specific places to meet multiple human needs in highly complex and often unexpected ways (Clark, 2007). It is from this highly complex and integrated place-based scenario of agricultural landscape sustainability that wicked problems emerge and wicked solutions are envisioned.

In the 1990s, Ostrom (2011) observed and envisioned self-governance solutions for the seeming impossibility of addressing the "tragedy of the commons" without relying on direct ownership or government regulations. In 2000, Costanza et al. moved the discussion beyond the "environment as a debate" and envisioned the need for an institutional framework to address the earth's renewable resources as assets to be managed and valued.

In 2002, Esty and Ivanova envisioned a network-based global environmental mechanism (GEM) containing data collection, calculations, and a repository for information to be used to assess environmental processes and trends, and assist forecasting long-term trends and environmental risks. The GEM would have a means to facilitate transactions related to environmental issues in return for payment or achieving policy objectives. These compelling visions have inspired countless sustainability efforts, including those discussed as case studies and the MSSG platform and eco-commerce ecosystem.

Abelow (2014) in *Imagining a Digital Future* envisions building a digitized earth to address many social issues. This includes a technological platform to connect farmers, manufacturers, retailers, health professionals, and consumers to figure out how to produce healthy food on sustainable farms. He refers to this process as new *digital governance*.

The vision of the earth as a digitized surface divided up into 6.4 trillion NCCs, with each representing a particular capacity and value associated with the economic system, is a compelling vision to achieve these objectives. Each of the NCCs produces a quantity of NCUs contributing to the health and wealth of the earth's inhabitants. The quantity and quality of the ecoservices they contribute is in direct proportion to how they are managed within the context of the climate and the environment. The MSSG platform allows one to envision the earth's landscape as the production floor of a living factory estimated to produce $124 T/year of ecoservices and goods: a total that exceeds the global GDP of $75 trillion/year. This startling comparison illustrates that the environment is not a subset of the economy, but the economy is a subset of the environment.

The vision of the eco-commerce ecosystem stakeholders is to add an *ecological dimension* to the economy. By supporting shared governance, it has the capacity to incorporate Ostrom's (2011) self-governance theories at glocal scales. It can transform environmental externalities that arise from open access regimes and undefined property rights into tradable ecoservices. It will allow multiple stakeholders to value agriculture landscape ecoservices through interdependent symbiotic transactions. Through this *symbiotic demand*, Porter and Kramer's (2011) corporate shared value can be realized from an environmental and sustainability perspective. The MSSG platform creates a business ecosystem that is able to capture the multiple flows of values from ecological ecosystems. An internal cryptocurrency and the unique transaction process enable the eco-commerce ecosystem participants to transcend the governance and economic issues associated with the "tragedy of the commons". It supports the emergence of a glocal ecosystem that simultaneously addresses the (1) various scopes and scales of natural capital outputs, (2) the disparate stakeholders and their varied accounting and valuation strategies, and (3) the conflicting governance styles of the stakeholder organizations.

It combines the low and diffuse values associated with ecoservices and reduces and shares the high transaction costs to resolve the enduring wicked problem associated with agriculture landscape sustainability.

chapter thirteen

Conclusion

As the world becomes more interconnected and glocalized, a greater number of society's problems will meet the criteria of a wicked problem. The complex systems of society, economies, and the environment are becoming increasingly intertwined and overwhelm the traditional problem-solving strategies of governments, institutions, corporations, and nongovernmental organizations (NGOs). Science and society need a transdisciplinary approach or they will continue to be debilitated by the potential of the next looming wicked problem. The emerging field of sustainability science needs to identify how to transcend the sector boundaries and better understand the principles of wicked problems.

13.1   Wicked principles

To take a different tack on the seven *sustainability science* questions asked by Kates (2010) in Box 1.2 is to ask *why* these questions are so difficult to answer or resolve. Analysis of the 11 agricultural landscape sustainability case studies revealed three principle sources of this wicked problem and why traditional problems solving does not work:

- Natural capital outputs and outcomes are varied in scope and scale.
- Agricultural stakeholders account for and value the landscape outputs and outcomes differently.
- Stakeholder organizational governance styles inherently conflict.

Since all wicked problems emerge from complex systems, and resolving wicked problems consists of changing the system as it *is* to as it *ought to be*, these three principle sources of agriculture's wicked problem may be viewed, generically, as three principle sources of wicked problems. As generic principles, these three causes may be common to *all* wicked problems:

- Outputs and outcomes of a system vary in scope and scale.
- System stakeholders have disparate values of and accounting processes for the outputs and outcomes of a system.
- Stakeholder organizations within a system have inherently conflicting governance styles.

In general terms, these three principles represent the science, economics, and social aspects of wicked problems.

13.2 Sustainability science

The central purpose of sustainability science is not just to analyze problems but to contribute to solving them (Hart and Bell, 2013). Sustainability science creates new knowledge, coproduced through close collaboration between scholars and practitioners (Clark, 2003). This broader understanding is gained from observations, monitoring, and place-based studies. It is intended to inspire practitioners and scientists to move scientific knowledge into action (Kates, 2010). Sustainability science solutions require a new space for interactions to occur.

For agricultural landscape sustainability, this new space was created by the platform's technological architecture. The resulting eco-commerce ecosystem acts as a *version of sustainability science*. It provides the space and intelligence to address Kate's (2010) sustainability questions. In the broadest of perspectives, the platform is based on a model representing the earth as a single ecological organism consisting of 6.4 billion natural capital cells (NCCs) supporting billions of interactions with billions of individuals and economic entities. To manage and sustain such a complex system one could only imagine a *DNA-type solution* could be intelligent, flexible, and robust enough to account for each NCC, the natural capital units (NCUs) it generates, and the exchanges that occur among billions of constituents.

13.3 A DNA solution

Biologists in the 1940s had difficulty in accepting DNA as the genetic material because of the apparent simplicity of its chemistry. DNA was known to be a long polymer composed of only four types of chemically similar subunits (Alberts et al., 2002). Despite evidence supporting DNA as the hereditary blueprint, scientists thought that the organizational structure needed to be *more* complicated.

Even with this relative simplicity of its core chemistry, DNA is not only the hereditary blueprint for life forms but provides the core functions of transcribing, translating, coding, and communicating. In many respects, DNA acts as a *smart* platform. It interacts with different molecules to enable a range of mechanisms affecting individual cells, organs, and/or the entire organism. The DNA molecule remains in the nucleus of the cells, but codes RNA strands (transcription) to transfer this information into the cell at-large to interact with enzymes (translation) and make proteins for specific functions. These proteins are modified and folded into structural and messenger proteins and enzymes and sent to the specific areas of the body, such as to the 13 organ systems and approximately 37 trillion cells.

This is all possible by imbuing everything with information, starting from the DNA molecule. The proteins that DNA encodes contain information to allow the organism to interact with its environment, which represents another source of information. Proteins and their products also interact and self-assemble to form organelles, tissues, organs, and organisms (Vincent et al., 2006).

Biologists have theorized that these DNA processes in complex multicellular organisms are accomplished through the combination of stable core processes at the cellular level, variable complementary processes that generate useful variation at the organism level, and a stable interface. This incredible process based on four amino acid subunits addresses what could be considered one of the most complex systems on the planet: life. Mimicking this biological process to resolve complex, wicked problems is a promising strategy.

13.4   Platform biomimicry

Biomimicry is the imitation of the models, systems, and mechanisms of nature for the purpose of solving human problems. Velcro is the archetype, inspired by the hooked seeds of burdock. In comparing human-engineered to biological systems, Vincent et al.'s (2006) research revealed human-engineered systems rely on controlling energy and substances to great degrees, where biological systems rely on information and structure. Depending on what system approach one mimics, the solution strategies would presumably yield much different results.

From this perspective, the multisided shared governance (MSSG) platform architecture mimics the DNA platform process that relies primarily on information and structure. Like the DNA platform, multisided platforms are composed of a stable core, a variable complementary component, and an interfacing system.

13.4.1   Stable core

The MSSG platform's core stable components consist of NCCs, NCC data, and indices. The NCCs are rigid delineations on the landscape and contain relatively static landscape data. The NCCs are analogous to the cells. The indices are analogous to the genetic codes of DNA molecules and represent what can be produced within the cell depending on internal and external signals. Landscape and management data are the subunits used to generate the "RNA" to transcribe the index score, export that information, and translate this information to other organizations. The recipients provide feedback to instigate or retard production of the variety of index values.

13.4.2   Variable complementary component

Like the DNA genetic code imbuing everything with information, a single index carries information that within the context of the particular recipient's objectives can be used to assess, plan, analyze, monitor, communicate, educate, research, and value outputs and outcomes.

Similar to the messenger role of RNA, the indices provide a continual source of information derived from the platform core. This relatively ubiquitous information enables self-organization among disparate stakeholders. It enables stakeholders to develop a variety of schemes to address government, NGO, utility, and corporate interests through the interdependent interactions among them. These schemes naturally seek the advantages of symbiotic demand, where sustainability constituents receive the information to substantiate their sustainability claims at the lowest cost possible.

In the cells, a similar exploratory process exists to provide organisms with an adaptation strategy to generate many different outcomes given a limited number of genes. For example, there is no predetermined genetic map for the distribution of blood vessels in the body, but the vascular system responds to signals from hypoxic tissues and generates the substances to produce blood vessels in the place and time needed (Kirschner and Gerhart, 1998).

13.4.3   Interface

The third component is the platform's interface where the stakeholders interact, exchange data, and conduct transactions. The importance of the interface and its stability is revealed in the evolution of computers. During the evolution of the original IBM PC, the relatively stable core processes were not conserved yet the interfaces that sit between the system board, the hardware, and the software were conserved (Baldwin and Woodard, 2009). Relative to the MSSG platform, this could mean the NCCs, indices, and NCUs are more prone to change than the interface.

13.5   A generic wicked solution platform

As the number of wicked problems increases, society needs to institute a new process to resolve them. The MSSG platform applied within the context of the incredible, but relatively simple DNA platform provides a potentially working example of a wicked solution model.

While the agriculture system as it *is* seems quite a distance from where this book describes where the system *ought to be*, there are many emerging components that will enable the transformation. The United States Department of Agriculture (USDA) is in its ninth decade of researching and applying technical solutions to natural resource problems. The corporate supply chain is undergoing extensive evaluation of its direct and indirect environmental footprint. Utilities are a recent participant in sustainability as they recognize the role of natural capital in supplying public needs, particularly drinking water, and standardization on how to account for its value. To provide the information and connect these disparate stakeholders, big data and multisided platforms have emerged as the basis for new business models. The ultimate test, of course, is generating an NCC-gridded earth and enabling a multisided platform to catalyze the interaction and the exchange of information and value.

Other wicked problems associated with human health, poverty, illicit drug use and trade, and climate may or may not have sufficient information to begin resolving them. Several of these are now referred to as super wicked problems. Super wicked problems contain all the wicked problem features and include exasperating aspects such as time is not costless, so the longer it takes to address the problem, the harder it will be to do so. A second issue is that those in the best position to address the problem are not only those who caused it, but also those with the least incentive to act. A third feature is the absence of an existing institutional framework of government with the ability to develop, implement, and maintain the laws necessary to address a problem (Lazarus, 2009).

Super wicked problems are indeed candidates for the simplicity and innovativeness embedded in a DNA-type platform. One strategy is to begin as life did. Simple systems (like bacteria) can and do vary their core processes (Kirschner and Gerhart, 1998). In both biological and economic systems, evolution proceeds via the mechanisms of variation and selective retention of advantageous forms (Campbell, 1965). The development of many complex multicellular organisms is based on *conservation* of core processes at the cellular level with a stable interface with complementary processes that support variation. This creates both flexible and robust mechanisms that support change and variability in other processes (Kirschner and Gerhart, 1998).

By reusing core components, the cost of variety and innovation at the system level can be reduced. The whole system does not have to be invented or rebuilt from scratch to generate a new product or respond to changes in the external environment. In this fashion, the platform system as a whole becomes evolvable: it can be adapted at low cost without losing its identity or continuity of design (Baldwin and Woodard, 2009).

13.6   A resolutionary path

Humanity has consistently faced seemingly insurmountable problems, often of their own creation from their behaviors, actions, and the characteristics of their existence: their DNA. These type of problems are often the most difficult to resolve as the search for the source of the problem leads back to humanity itself. There are few things more challenging than recognizing and admitting that the challenging issue at hand emerged from one's persona, community, and culture.

Resolving wicked problems requires one to address the problem at the beginning of its life cycle where it emerges, rather than at the end of its life cycle where it manifests itself. It takes keen observation skills to recognize the quiet source of these problems amid the obvious results. And it takes even more tact and skill to convince others.

The solutions require shifts in paradigms, not continuous improvements of the familiar processes. The system as it *is* must somehow be transformed into the system as it *ought to be* through the participants themselves transcending their traditional perceptions and perspectives. It is the participants currently vested in the system and creating the resiliency of the system at hand that are able to change it. To change such a system from a social perspective, the culture and governance of society and its institutions must evolve to resolve wicked problems. The platform from which society and institutions emerge and obtain their resiliency must emerge and evolve.

Bibliography

103rd Congress. Public Law 103. *Title I—Federal Crop Insurance*, October 13, 1994. Accessed April 29, 2016. https://www.congress.gov/bill/103rd-congress/house-bill/4217.

Abdalla, C. Land use policy: Lessons from water quality markets. *Choices*, 23, 4th Quarter (2008): 22–28.

Abdalla, C., T. Borisova, D. Parker, and K. S. Blunk. Water quality credit trading and agriculture: Recognizing the challenges and policy issues ahead. *Choices* 22, no. 2 (Spring 2007): 117–24.

Abelow, D. *Imagining a Great Future: Creating Greatness for All.* Dan Abelow, 2014. http://www.amazon.com/dp/B00IPYYJMI/ref=rdr_kindle_ext_tmb

Adner, R. Match your innovation strategy to your innovation ecosystem. *Harvard Business Review* 84, no. 4 (April 2006): 98–107.

Alberts, B., A. Johnson, and J. Lewis. *Molecular Biology of the Cell.* 4th ed. New York: Garland Science, 2002.

Alcamo, J., and E. M. Bennett. *Ecosystems and Human Well-being: A Framework for Assessment.* Washington, DC: Island Press, 2003.

Alonso, J. M., J. Clifton, and D. Díaz-Fuentes. Did New Public Management Matter? An empirical analysis of the outsourcing and decentralization effects on public sector size. *Public Management Review* 17, no. 5 (2013): 643–60. doi:10.1080/14719037.2013.822532.

Amor, D. *The E-business (R)evolution: Living and Working in an Interconnected World.* Upper Saddle River, NJ: Prentice Hall PTR, 2000.

Andonova, L. B. Public-private partnerships for the earth: Politics and patterns of hybrid authority in the multilateral system. *Global Environmental Politics* 10, no. 2 (2010): 25–53. doi:10.1162/glep.2010.10.2.25.

APWA. Strategic Plan. APWA—Center for Sustainability. 2013. Accessed January 02, 2016. http://www.apwa.net/centerforsustainability/elements.

Armstrong, M. Competition in two-sided markets. *The RAND Journal of Economics* 37, no. 3 (2006): 668–91. doi:10.1111/j.1756-2171.2006.tb00037.x.

Avison, D. E., P. A. Golder, and H. U. Shah. Towards an SSM toolkit: Rich picture diagramming. *European Journal of Information Systems* 1, no. 6 (1992): 397–408. doi:10.1057/ejis.1992.17.

Avril, E., and C. Zumello. Introduction: Towards organizational democracy? Convergence and divergence in models of economic and political governance. In *New Technology, Organizational Change, and Governance*, edited by E. Avril, and C. Zumello. UK: Palgrave Macmillan, 2013, 1–20.

Axelrod, R. M., and M. D. Cohen. *Harnessing Complexity: Organizational Implications of a Scientific Frontier.* New York: Free Press, 1999.

Baghbadorani, M., and A. Harandi. A conceptual model for business ecosystem and implications for future research. *International Proceedings of Economics Development and Research* 52, no. 17 (2012): 82–86.

Baldwin, C., and C. J. Woodard. The architecture of platforms: A unified view. In *Platforms, Markets, and Innovation,* edited by A. Gawer. Cheltenham, UK: Edward Elgar, 2009.

Barnhart, Robert K., and Sol Steinmetz. *The Barnhart Dictionary of Etymology.* Bronx, NY: H.W. Wilson, 1988. Balint, P. J. *Wicked Environmental Problems: Managing Uncertainty and Conflict.* Washington, DC: Island Press, 2011.

Bell, S., and S. Morse. *Sustainability Indicators: Measuring the Immeasurable?* London, UK: Earthscan, 2008.

Bennett, K. Additionality: The next step for ecosystem service markets. *Duke Environmental Law and Policy Forum* 20 (Summer 2010): 417–38.

Bennun, L. A., J. Ekstrom, and J. Bull. *Integrating the Value of Natural Capital into Private and Public Investment: The Role of Information.* Cambridge, UK: Biodiversity Consultancy, 2014.

Berge, E., and F. Van Laerhoven. Governing the commons for two decades: A complex story. *International Journal of the Commons* 5, no. 2 (2011): 160. doi:10.18352/ijc.325.

Bevir, M. *Key Concepts in Governance.* Los Angeles, CA: SAGE, 2009.

Bhan, M. The role of public-private partnerships in U.S. environmental policy: Case of the EPA and the U.S. semiconductor industry. *The Public Purpose* XI (2013): 49–68.

Blatter, J. K. *Encyclopedia of Governance.* Thousand Oaks, CA: Sage Online, 2007.

Bloom, L., and M. Lahey. *Language Development and Language Disorders.* New York: Wiley, 1978.

Bollier, D., and S. Helfrich. *The Wealth of the Commons: A World beyond Market and State.* Amherst, MA: Levellers Press, 2012.

Boyd, J., and S. Banzhaf. What are ecosystem services? The need for standardized environmental accounting units. *Ecological Economics* 63, no. 2–3 (2007): 616–26. doi:10.1016/j.ecolecon.2007.01.002.

Brandt, P., A. Ernst, F. Gralla, C. Luederitz, D. J. Lang, F. Newig, F. Reinert, D. J. Abson, and H. Von Wehrden. A review of transdisciplinary research in sustainability science. *Ecological Economics* 92 (2013): 1–15. doi:10.1016/j.ecolecon.2013.04.008.

Brown, V. A. *Tackling Wicked Problems: Through the Transdisciplinary Imagination.* London, UK: Earthscan, 2010.

Brown, T. C., J. Bergstrom, and J. Loomis. Defining, valuing, and providing ecosystem goods and services. *Natural Resources Journal* 47 (2007): 329–76.

Bruggemann, J., M. Rodier, M. Guillaume, and S. Andréfouët. Wicked social-ecological problems forcing unprecedented change on the latitudinal margins of coral reefs: The case of southwest Madagascar. *Ecology and Society E#x0026;S* 17, no. 4 (2012): 47. doi:10.5751/es-05300-170447.

Börzel, T., S. Guttenbrunner, and S. Seper. *Conceptualizing New Modes of Governance in EU Enlargement.* Working paper no. Project Description. Vol. 12/D1. CIT1-CT-2004-506392. Free University of Berlin, 2005. Accessed April 29, 2016. http://userpage.fu-berlin.de/~europe/forschung/eu/12d1.pdf.

Buizer, M., B. Arts, and K. Kok. Governance, scale and the environment: The importance of recognizing knowledge claims in transdisciplinary arenas. *Ecology and Society*, 21st ser., 16, no. 1 (2011): 21. doi:10.1177/0263774x15614725.

Burgos, S. Corporations and social responsibility: NGOs in the ascendancy. *Journal of Business Strategy* 34, no. 1 (2012): 21–29. doi:10.1108/02756661311301756.

Camillus, J. Strategy as a wicked problem. *Harvard Business Review* 86, no. 5 (May 2008): 98–106.

Campbell, D. Variation and selective retention in socio-cultural evolution. In *Social Change in Developing Areas; A Reinterpretation of Evolutionary Theory*, edited by H. R. Barringer, G. I. Blanksten and R. W. Mack. Cambridge, MA: Schenkman Publisher, 1965, 19–49.

Caris, N., H. Underwood, and N. A. Campbell. *Instructor's Guide for Campbell's Biology.* 4th ed. Menlo Park, CA: Benjamin/Cummings, 1996.

Carpenter, S., B. Walker, J. M. Anderies, and N. Abel. From metaphor to measurement: Resilience of what to what? *Ecosystems* 4, no. 8 (2001): 765–81. doi:10.1007/s10021-001-0045-9.

CASS. Conference of the Swiss Scientific Academies *Research on Sustainability and Global Change: Visions in Science Policy by Swiss Researchers.* Bern, Switzerland: ProClim-Swiss Academy of Sciences, 1997.

Castells, M. *The Information Age: Economy, Society and Culture.* Vol. 1. Oxford, UK: Blackwell Publ., 1996.

CBP. *Strengthening Verification of Best Management Practices Implemented in the Chesapeake Bay Watershed: A Basinwide Framework.* Report. Annapolis, MD: CBP, October 2014.

Chaffin, B. C., H. Gosnell, and B. A. Cosens. A decade of adaptive governance scholarship: Synthesis and future directions. *Ecology and Society* 19, no. 3 (2014): 56. doi:10.5751/es-06824-190356.

Cherry, S., T. J. Canfield, and M. McNeil. *Wetlands and Water Quality Trading: Review of Current Science and Economic Practices with Selected Case Studies.* Ada, OK: Ground Water and Ecosystems Restoration Division, National Risk Management Laboratory, Office of Research and Development, 2007.

Choi, B., and A. Pak. Multidisciplinarity, interdisciplinarity and transdisciplinarity in health research, services, education and policy. *Clinical of Investigative Medicine* 29, no. 6 (2006): 351–64.

Choudary, S. P. *Platform Power: Secrets of Billion Dollar Internet Start-ups.* S. P. Choudary, 2013.

Choudary, S. Paul, G. Parker, and M. Van Alystne. *Platform Scale: How an Emerging Business Model Helps Startups Build Large Empires with Minimum Investment.* Platform Thinking Labs, 2015.

Christensen, C. M. *The Innovator's Dilemma: When New Technologies Cause Great Firms to Fail.* Boston, MA: Harvard Business School Press, 1997.

Christensen, K. Building Shared Understanding of Wicked Problems. Interview *Rotman Magazine*, Winter 2009, 17–20. Accessed April 29, 2016. http://www.cognexus.org/Rotman-interview_SharedUnderstanding.pdf.

Christensen, N. L., A. M. Bartuska, J. H. Brown, S. Carpenter, C. D'antonio, R. Francis, J. F. Franklin, J. A. Macmahon, R. F. Noss, D. J. Parsons, C. H. Peterson, M. G. Turner, and R. G. Woodmansee. The report of the ecological society of America committee on the scientific basis for ecosystem management. *Ecological Applications* 6, no. 3 (1996): 665. doi:10.2307/2269460.

Christensen, T., and P. Lægreid. Democracy and administrative policy: Contrasting elements of new public management (NPM) and post-NPM. *European Political Science Review* 3, no. 01 (2011): 125–46. doi:10.1017/s1755773910000299.

Clark, J. G. Economic development vs sustainable societies: Reflections on the players in a crucial contest. *Annual Review of Ecology and Systematics* 26, no. 1 (1995): 225–48. doi:10.1146/annurev.ecolsys.26.1.225.

Clark, W. C. Sustainability science: A room of its own. *Proceedings of the National Academy of Sciences* 104, no. 6 (2007): 1737–738. doi:10.1073/pnas.0611291104.

Clark, W. C., and N. M. Dickson. Sustainability science: The emerging research program. *Proceedings of the National Academy of Sciences* 100, no. 14 (2003): 8059–061. doi:10.1073/pnas.1231333100.

Claussen, E. Global environmental governance issues for the new US administration. *Environment: Science and Policy for Sustainable Development* 43, no. 1 (2001): 28–34.

Clayton, C. Noting efforts to grow water quality trading programs. DTN/The Progressive Farmer. September 17, 2015. Accessed January 04, 2016. http://www.dtnprogressivefarmer.com.

Clayworth, J. Cities often send nitrates downstream after treatment. Des Moines Register. June 14, 2014. Accessed January 02, 2016. http://www.desmoinesregister.com/story/news/2014/06/15/cities-often-send-nitrates-downstream-treatment/10544605/.

Coase, R. H., O. E. Williamson, and S. G. Winter. *The Nature of the Firm*. New York: Oxford University Press, 1993.

Commonwealth of Australia 2007. *Tackling Wicked Problems: A Public Policy Perspective*. Canberra, Australia: Australian Public Service Commission, 2007.

Conklin, E. J. *Dialogue Mapping: Building Shared Understanding of Wicked Problems*. Chichester, England: Wiley, 2006.

Cooley, D., and L. Olander. *Stacking Ecosystem Services Payments Risks and Solutions*. Working paper no. NI WP 11-04. Nicholas Institute for Environmental Policy Solutions, 2011.Accessed April 29, 2016. https://nicholasinstitute.duke.edu/ecosystem/land/stacking-ecosystem-services-payments

Cosman, D., R. Schmidt, J. Harrison-Cox, and D. Batker. How water utilities can spearhead natural capital accounting. *Solutions* 2, no. 6 (January 2012): 28–31.

Costanza, R., R. D'arge, R. De Groot, S. Farber, M. Grasso, B. Hannon, K. Limburg, S. Naeem, R. V. O'neill, J. Paruelo, R. G. Raskin, P. Sutton, and M. Van Den Belt. The value of the world's ecosystem services and natural capital. *Nature* 387, no. 6630 (1997): 253–60. doi:10.1038/387253a0.

Costanza, Robert, Herman Daly, Carl Folke, Paul Hawken, C.S. Holling, Anthony McMichael, David Pimental, and David Rapport. "Managing Our Environmental Portfolio. *BioScience* 50, no. 2 (February 01, 2000): 149-55.

Costanza, R., R. De Groot, P. Sutton, S. Van Der Ploeg, S. J. Anderson, I. Kubiszewski, S. Farber, and R. K. Turner. Changes in the global value of ecosystem services. *Global Environmental Change* 26 (2014): 152–58. doi:10.1016/j.gloenvcha.2014.04.002.

Covey, S. M. R., and R. R. Merrill. *The Speed of Trust: The One Thing That Changes Everything*. New York: Free Press, 2006.

Cox, C. Conservation Intelligence. *Journal of Soil and Water Conservation* 60, no. 3 (2005): 50A–4A.

Cvetkovic, M., and P. Chow-Fraser. Use of ecological indicators to assess the quality of Great Lakes coastal wetlands. *Ecological Indicators* 11, no. 6 (2011): 1609–622. doi:10.1016/j.ecolind.2011.04.005.

Daily, G. C. What are ecosystem services? Introduction to *Nature's Services: Societal Dependence on Natural Ecosystems,* edited by G. C. Daily. Washington, DC: Island Press, 1997.

Davies, K. K., K. T. Fisher, M. E. Dickson, S. F. Thrush, and R. Le Heron. Improving ecosystem service frameworks to address wicked problems. *Ecology and Society* 20, no. 2 (2015): 37. doi:10.5751/es-07581-200237.

Davis, S. M. *Managing Corporate Culture.* Cambridge, MA: Ballinger, 1984.

De Loe, R., D. Armitage, R. Plummer, S. Davidson, and L. Moraru. *From Government to Governance: A State-of-the-Art Review of Environmental Governance.* Report no. 2009/ES-001 Final Report. Guelph, ON: Prepared for Alberta Environment, Environmental Stewardship, Environmental Relations: Rob De Loë Consulting Services, 2009.

Defries, R., G. P. Asner, and J. Foley. A glimpse out the window: Landscapes, livelihoods, and the environment. *Environment: Science and Policy for Sustainable Development* 48, no. 8 (2006): 22–36. doi:10.3200/envt.48.8.22-36.

Dietz, T. The struggle to govern the commons. *Science* 302, no. 5652 (2003): 1907–912. doi:10.1126/science.1091015.

Dijk, J. V., and A. Winters-van Beek. The perspective of network government: The struggle between hierarchies, markets and networks as modes of governance in contemporary government. In *ICTs, Citizens and Governance: After the Hype!,* edited by A. Meijer, K. Boersma and P. Wagenaar, 235–55. Amsterdam, the Netherlands: IOS Press, 2009.

DMWW. Des Moines Water Works. 2015. Accessed January 02, 2016. http://www.dmww.com/.

Donahue, J. D., and J. S. Nye. *Market-Based Governance: Supply Side, Demand Side, Upside, and Downside.* Cambridge, MA: Brookings Institution Press, 2002.

EDF. EDF Launches initiative to reduce fertilizer pollution from commodity grain crops. News release, October 8, 2014. Environmental Defense Fund. Accessed January 02, 2016. http://www.edf.org/media/edf-launches-initiative-reduce-fertilizer-pollution-commodity-grain-crops.

Eggers, W., and A. Muoio. Wicked opportunities. In *Business Ecosystems Come of Age,* edited by E. Kelly, 31–42. Business Trends. Deloitte University Press, 2015.

EPA. *Water on Tap: What You Need to Know.* Washington, DC: U.S. Environmental Protection Agency, Office of Water, 2009. http://nepis.epa.gov/Exe/ZyPDF.cgi/P1008ZP0.PDF?Dockey=P1008ZP0.PDF.

EPA. *Planning for Sustainability: A Handbook for Water and Wastewater Utilities.* Vol. EPA-832-R-12-001. Washington DC: EPA, 2012.

EPA. National summary of impaired waters and TMDL information. Iaspubepa.gov. 2015. http://iaspub.epa.gov/waters10/attains_nation_cy.control?p_report_type=T#APRTMDLS. last accessed 4/29/16

EPRI. Pilot trading plan 1.0 for the Ohio River Basin interstate water quality trading project. Project Description. August 1, 2012. http://wqt.epri.com/pdf/Full-Trading-Plan-as-amended.pdf.

Esty, D. C., and M. H. Ivanova. *Global Environmental Governance: Options and Opportunities*. New Haven, CT: Yale School of Forestry and Environmental Studies, 2002.

Evans, D. S. *Platform Economics: Essays on Multi-Sided Businesses*. 2011. https://www.competitionpolicyinternational.com/about-us/

Evans, D. S., and R. Schmalensee. *The Antitrust Analysis of Multi-Sided Platform Businesses*. Working paper no. 623. Coase-Sandor Institute for Law and Economics, 2012. Accessed April 29, 2016. http://www.nber.org/papers/w18783 4/29/16

Falkner, R. Private environmental governance and international relations: Exploring the links. *Global Environmental Politics* 3, no. 2 (2003): 72–87. doi:10.1162/152638003322068227.

Farmer, J. D., M. Gallegati, C. Hommes, A. Kirman, P. Ormerod, S. Cincotti, A. Sanchez, and D. Helbing. A complex systems approach to constructing better models for managing financial markets and the economy. *The European Physical Journal Special Topics* 214, no. 1 (2012): 295–324. doi:10.1140/epjst/e2012-01696-9.

Fábián, A. New public management and what comes after. *Issues of Business and Law* 2, no. -1 (2010): 36–45. doi:10.2478/v10088-010-0004-y.

Feldman, I. R. Business and industry: Transitioning to sustainability. In *Agenda for a Sustainable America*, edited by J. C. Dernbach, 71–91. Washington, DC: ELI Press, Environmental Law Institute, 2009.

Fenichel, E. P., and J. K. Abbott. Natural capital: From metaphor to measurement. *Journal of the Association of Environmental and Resource Economists* 1, no. 1/2 (2014): 1–27. doi:10.1086/676034.

Field to Market. The alliance for sustainable agriculture. FieldtoMarket.org. December 11, 2014a. Accessed April 29, 2016. https://www.fieldtomarket.org/report/national-2/FINAL_Fact_Sheet-What_is_Field_to_Market_121114.pdf.

Field to Market. Field to Market and the Sustainability Consortium announce partnership to harmonize measurement and reporting of sustainable agriculture. News release, November 21, 2014b. Accessed April 29, 2016. https://www.fieldtomarket.org/news/2014/field-to-market-and-the-sustainability-consortium-announce-partnership-to-harmonize-measurement-and-reporting-of-sustainable-agriculture/.

Fisher, D. A. *An Emergent Perspective on Interoperation in Systems of Systems*. Technical paper no. ESC-TR-2006-003. Pittsburgh, PA: Carnegie Mellon University, 2006.

Fleming, J., and R. Rhodes. *It's Situational: The Dilemmas of Police Governance in the 21st Century*. Proceedings of Australasian Political Studies Association Conference, University of Adelaide, Australia. 2004.

Folke, C., T. Hahn, P. Olsson, and J. Norberg. Adaptive goverance of social-ecological systems. *Annual Review Environment Resources* 30 (2005): 441–73.

Fox, Jessica. "Ohio River Basin Trading Project." EPRI, January 2016, 1.

Frederickson, H. G. *Whatever Happened to Public Administration? Governance, Governance Everywhere*. Working paper no. QU/GOV/3/2004. Belfast: Institute of Governance Public Policy and Social Research, 2004.

Friedman, M. *Market Mechanisms and Central Economic Planning*. Lecture, G. Warren Nutter Lecture in Political Economy. Stanford, CA: Hoover Institution, March 4, 1981.

Friedman, T. L. Advice for China. The New York Times. June 04, 2011. Accessed January 02, 2016. http://www.nytimes.com/2011/06/05/opinion/05friedman. html?_r=0.

Friedman, T. L. Stampeding black elephants. The New York Times. November 22, 2014. Accessed January 04, 2016. http://www.nytimes.com/2014/11/23/ opinion/sunday/thomas-l-friedman-stampeding-black-elephants. html?_r=1.

FtoM. Field To Market: The Alliance for Sustainable Agriculture. 2016. Accessed January 03, 2016. https://www.fieldtomarket.org/.

Fulton, G. Towards Landscape Intelligence. *Landscape Urbanism* 31 (April 2011): 46–53.

GASB. Financial accounting standards board. GASB: Mission, Vision, and Core Values. 2015. Accessed January 02, 2016. http://www.gasb.org/jsp/GASB/ Page/GASBSectionPage&cid=1175804850352.

Gieseke, T. *Livestock Environmental Quality Assurance Program.* Report. St. Paul, MN: Minnesota Dept. of Agriculture, 2011.

Gieseke, T. M. *EcoCommerce 101: Adding an Ecological Dimension to the Economy.* Minneapolis, MN: Bascom Hill Publishing Group, 2011.

Glouberman, S., and B. Zimmerman. *Complicated and Complex Systems: What Would Successful Reform of Medicare Look Like?* Working paper no. 8. Commission on the Future of Health Care in Canada, 2002. file:///C:/Users/Tim%20 Gieseke/Downloads/complicatedandcomplexsystems-zimmermanreport- medicare-reform.pdf 4/29/16

Godsiff, P. Bitcoin: Bubble or blockchain. *Agent and Multi-Agent Systems: Technologies and Applications Smart Innovation, Systems and Technologies.* Springer International Publishing, 2015, 191–203. doi: 10.1007/978-3-319-19728-9_16.

González-Gaudiano, E. Glocalization and Sustainability. *Critical Forum* 5 (August 1997): 83-90. Accessed April 29, 2016. http://www.anea.org.mx/ glocalization-and-ustainability/.

Goodwin, T. The battle is for the customer interface. *Techcrunch.com*, March 3, 2015.

Govindarajan, V., and A. K. Gupta. Strategic innovation: A conceptual road map. *Business Horizons* 44, no. 4 (2001): 3–12. doi:10.1016/s0007-6813(01)80041-0.

Gray, B. Enhancing transdisciplinary research through collaborative leadership. *American Journal of Preventive Medicine* 35, no. 2 (2008): S124–32. doi:10.1016/j. amepre.2008.03.037.

The Guardian. Prince Charles tells business leaders that nature's 'bank' is being depleted. Guardian.com. December 13, 2013. Accessed April 29, 2016. http://www.theguardian.com/environment/2013/dec/13/prince- charles-business-leaders-nature-bank.

Gunderson, L. Resilience, flexibility and adaptive management antidotes for spurious certitude? *Conservation Ecology*, 7th ser., 3, no. 1 (1999): 7. http://www. consecol.org/vol3/iss1/art7/.

Gupta, J. Glocal water governance. *Water for a Changing World—Developing Local Knowledge and Capacity.* Proceedings of the International Symposium on Water for a Changing World—Developing Local Knowledge and Capacity. Delft, The Netherlands, June 13–15, 2007, 2008.

Gupta, J., C. Pahl-Wostl, and R. Zondervan. 'Glocal' water governance: A multi- level challenge in the anthropocene. *Current Opinion in Environmental Sustainability* 5, no. 6 (2013): 573–80. doi:10.1016/j.cosust.2013.09.003.

Gwartney, J. D. *Common Sense Economics: What Everyone Should Know about Wealth and Prosperity.* New York: St. Martin's Press, 2010.

Hadorn, G. H., D. Bradley, C. Pohl, S. Rist, and U. Wiesmann. Implications of transdisciplinarity for sustainability research. *Ecological Economics* 60, no. 1 (2006): 119–28. doi:10.1016/j.ecolecon.2005.12.002.

Hagel, J. The power of platforms. In *Business Ecosystems Come of Age*, edited by E. Kelly, 79–90. Deloitte University Press, 2015. http://dupress.com/about-deloitte-university-press/

Hagiu, A. Strategic decisions for multisided platforms. *MIT Sloan*, Winter 2014. http://sloanreview.mit.edu/article/strategic-decisions-for-multisided-platforms/. http://hbswk.hbs.edu/item/multi-sided-platforms2 4/29/16

Hagiu, A., and J. Wright. *Multi-Sided Platforms.* Working paper no. 12-024. 2011. http://hbswk.hbs.edu/item/multi-sided-platforms2 4/29/16

Haines-Young, R. H., and M. B. Potschin. *Methodologies for Defining and Assessing Ecosystem Services.* Technical paper no. C08-0170-0062. Final Report ed. JNCC, 2009. http://www.nottingham.ac.uk/cem/pdf/JNCC_Review_Final_051109.pdf 4/29/16

Halldorsson, G., B. D. Sigurdsson, L. Finér, J. Gudmundsson, T. Kätterer, B. R. Singh, L. Vesterdal, and A. Arnalds. Soil carbon sequestration—For climate, food security and ecosystem services. *Norden*, 2015. doi:10.6027/anp2015-792. pages 6–15

Hardin, G. Tragedy of the commons. *Science* 162, no. 3859 (December 1968): 1243–248. doi:10.1126/science.162.3859.1243.

Harrison, A. Environmental issues and the SNA. *Review of Income and Wealth* 35, no. 4 (1989): 377–88. doi:10.1111/j.1475-4991.1989.tb00599.x.

Hart, D., and K. Bell. Sustainability science: A call to collaborative action. *Agricultural and Resource Economics Review* 42, no. 1 (April 2013): 75–89.

Hart, D. D., K. P. Bell, L. A. Lindenfeld, S. Jain, T. R. Johnson, D. Ranco, and B. Mcgill. Strengthening the role of universities in addressing sustainability challenges: The Mitchell Center for Sustainability Solutions as an institutional experiment. *Ecology and Society E&S* 20, no. 2 (2015): 4. doi:10.5751/es-07283-200204.

Hartel, D. *2003 National Urban Forest Conference: Urban Natural Resources as Capital Assets.* GASB 34. September 17–20, 2003. http://www.urbanforestrysouth.org/resources/library/citations/Citation.2004-11-03.P194/fss_get/file.

Hatfield-Dodds, S., R. Nelson, and D. Cook. *Adaptive Governance: An Introduction, and Implications for Public Policy.* Proceedings of 51st Annual Conference of the Australian Agricultural and Resource Economics Society. Queenstown, NZ, February 13–16, 2007.

Hearnshaw, E., J.-M. Tompkins, and R. Cullen. *Addressing the Wicked Problem of Water Resource Management: An Ecosystem Services Approach.* Working paper. Melbourne, Australia: 55th Annual AARES National Conference, 2011.

Hejnowicz, A. P., D. G. Raffaelli, M. A. Rudd, and P. C. L. White. Evaluating the outcomes of payments for ecosystem services programs using a capital asset framework. *Ecosystem Services* 9 (2014): 83–97. doi:10.1016/j.ecoser.2014.05.001.

Helms, D. Hugh Hammond Bennett and the creation of the soil erosion service. *Journal of Soil and Water Conservation* 64, no. 2 (2009): 68A–74A. doi:10.2489/jswc.64.2.68a.

Herbst, P. G. *Alternatives to Hierarchies.* Leiden: M. Nijhoff Social Sciences Division, 1976.

Heywood, A. *Politics*. New York: Palgrave, 2003.

Ho, T. V. T., S. Woodley, A. Cottrell, and P. Valentine. A multilevel analytical framework for more-effective governance in human-natural systems: A case study of marine protected areas in Vietnam. *Ocean and Coastal Management* 90 (2014): 11–19. doi:10.1016/j.ocecoaman.2013.12.015.

Hodge, G. A., and C. Greve. *The Challenge of Public-Private Partnerships: Learning from International Experience*. Cheltenham, UK: Edward Elgar Pub., 2005.

Howes, S., and P. Wyrwoll. *Asia's Wicked Environmental Problems*. ADBI Working Paper 348. Tokyo, Japan: Asian Development Bank Institute, February 2012. doi:10.2139/ssrn.2013762.

Hunt, F., and S. Thornsbury. Facilitating transdisciplinary research in an evolving approach to science. *JSS Open Journal of Social Sciences* 02, no. 04 (2014): 340–51. doi:10.4236/jss.2014.24038.

Huppé, G., H. Creech, and D. Knoblauch. *The Frontiers of Networked Governance*. Working paper. Winnipeg: International Institute for Sustainable Development, 2012.

Hurst, K. The science behind the sustainability consortium. GreenBiz. April 4, 2014. Accessed January 03, 2016. http://www.greenbiz.com/video/2014/04/04/greenbiz-forum-2014-sustainability-consortium.

Jackson, N. *Tackling the Wicked Problem of Creativity in Higher Education*. Proceedings of ARC Centre for the Creative Industries and Innovation, International Conference Brisbane, June 2008. Guildford, England: Surrey Centre for Excellence in Professional Training and Education.

Jager, J. Foreword. In *Handbook of Transdisciplinary Research*, edited by G. H. Hadorn. Dordrecht, Netherlands: Springer, 2008.

Jefferies, M. Exxon Valdez oil spill. Anew NZ. Accessed January 03, 2016. http://www.anewnz.org.nz/page.asp?id=3228879138356TLQ.

Jessop, B. Governance and meta-governance in the face of complexity: On the roles of requisite variety, reflexive observation, and romantic irony in participatory governance. In *Participatory Governance in Multi-Level Context*, edited by H. Heinelt, 33–58, 2002. doi:10.1007/978-3-663-11005-7_2.

Jessop, B. *Governance, Governance Failure and Meta-Governance*. Proceedings of Policies, Governance and Innovation for Rural Areas, International Seminar November 21–23, 2003. Universita della Calabria, Arcavacata di Rende

Jessop, B. *The Governance of Complexity and the Complexity of Governance, Revisited*. Proceedings of Complexity, Science and Society Conference. Liverpool, 2005. September 11-14

Johnson, R., and J. Monke. *What Is the Farm Bill?* Issue brief no. 7-5700. Washington, DC: Congressional Research Service, 2014.

Jones, H. *Taking Responsibility for Complexity: How Implementation Can Achieve Results in the Face of Complex Problems*. Working paper no. 330. London: Overseas Development Institute, 2011.

Kamarck, E. C. *The End of Government. . .as We Know It: Making Public Policy Work*. Boulder, CO: Lynne Rienner Publishers, 2007.

Kates, R. Readings in sustainability science and technology. CID Working Paper No. 213. *Center for International Development*, Cambridge, MA: Harvard University, December 2010, 1–44.

Kates, R. W. What kind of a science is sustainability science? *Proceedings of the National Academy of Sciences* 108, no. 49 (2011): 19449–9450. doi:10.1073/pnas.1116097108.

Kates, R. W., W. Clark, R. Corell, and J. Hall. Environment and development: Sustainability science. *Science* 292, no. 5517 (2001): 641–42. doi:10.1126/science.1059386.

Kelly, E. *Business Ecosystems Come of Age.* Business Trends. Deloitte University Press, 2015. http://dupress.com/about-deloitte-university-press/

Kersbergen, K. V., and F. Van Waarden. 'Governance' as a bridge between disciplines: Cross-disciplinary inspiration regarding shifts in governance and problems of governability, accountability and legitimacy. *European Journal of Political Research* 43, no. 2 (2004): 143–71. doi:10.1111/j.1475-6765.2004.00149.x.

Kim, O., M. Walker, and W. S. Dawes. *The Free Rider Problem: Experimental Evidence.* Stony Brook, NY: Economic Research Bureau, State University of New York at Stony Brook, 1980.

Kimble, M. *Request for contribution.* E-mail message to author. July 8, 2014.

King, D., and P. Kuch. Will Nutrient Credit Trading Ever Work? An Assessment of Supply and Demand Problems and Institutional Obstacles. *ELR News and Analysis*, May 2003, 10352–0368.

Kirschner, Marc, and John Gerhart. "Evolvability." Proceedings of the National Academy of Sciences of the United States of America 95, no. 15 (July 21, 1998): 8420-427. Accessed April 29, 2016. http://www.jstor.org/stable/10.2307/45773?ref=search-gateway:7df137c7c24a9e384301d05082e748b2.

Klijn, E. H., and J. F. M. Koppenjan. Public management and policy networks: Foundations of a network approach to governance. *Public Management* 2, no. 2 (2000): 135–58. doi:10.1080/146166700411201.

Kontogianni, A., G. W. Luck, and M. Skourtos. Valuing ecosystem services on the basis of service-providing units: A potential approach to address the 'Endpoint Problem' and improve stated preference methods. *Ecological Economics* 69, no. 7 (2010): 1479–487. doi:10.1016/j.ecolecon.2010.02.019.

Kroemer, J. Meta-governors and Their Influence on Network Functioning: A Study of Meta-governance in the Case of the European City Network Eurocities. Master's thesis, Erasmus University Rotterdam, Rotterdam, The Netherlands, 2010.

Ladyman, J., and J. Lambert. *What Is a Complex System?* Publication. Bristol, UK: Department of Mathematics and Centre for Complexity Sciences, University of Bristol, 2012.

Laloux, F. *Reinventing Organizations: A Guide to Creating Organizations Inspired by the Next Stage in Human.* Salt Lake City, UT: Nelson Parker, 2014.

Landell-Mills, N., and I. T. Porras. *Silver Bullet or Fools' Gold?: A Global Review of Markets for Forest Environmental Services and Their Impact on the Poor.* London, UK: IIED, 2002.

Landers, D. H., and A. M. Nahlik. *Final Ecosystem Goods and Services Classification System (FEGS-CS).* Publication no. EPA/600/R-13/ORD-004914. Washington, DC: U.S. Environmental Protection Agency, Office of Research and Development, 2013.

Lang, D. J., A. Wiek, M. Bergmann, M. Stauffacher, P. Martens, P. Moll, M. Swilling, and C. J. Thomas. Transdisciplinary research in sustainability science. *Sustainability Science* 7, no. S1 (2012): 25–43. doi:10.1007/s11625-011-0149-x.

Lazarus, R. J. Super wicked problems and climate change: Restraining the present to liberate the future. *Cornell Law Review* 94 (2009): 1155–231.

LCCMR. Environmental practices on dairy farms subd. 08c. M.L. 2001 Projects. July 1, 2001. Accessed January 03, 2016. http://www.lccmr.leg.mn/projects/2001-index.html#20018c.

Lemos, M. C., and A. Agrawal. Environmental governance. *Annual Review of Environment and Resources* 31, no. 1 (2006): 297–325. doi:10.1146/annurev.energy.31.042605.135621.

Libecap, G. D. The tragedy of the commons: Property rights and markets as solutions to resource and environmental problems. *Australian Journal of Agricultural and Resource Economics* 53, no. 1 (2009): 129–44. doi:10.1111/j.1467-8489.2007.00425.x.

Lihua, H., G. Hu, and X. Lu. E-business ecosystem and its evolutionary path: The case of the Alibaba. *Pacific Asia Journal of the Association for Information Systems*, 3rd ser., 1, no. 4 (2010): 1–12. http://aisel.aisnet.org/pajais/vol1/iss4/3/. page

Liu, H., Z. Tian, and X. Guan. Analysis on complexity and evolution of E-commerce ecosystem. *International Journal of U- and E-Service, Science and Technology* 6, no. 6 (2013): 41–50. doi:10.14257/ijunesst.2013.6.6.05.

Lloyd, W. F. *Two Lectures on the Checks to Population Delivered before the University of Oxford, in Michaelmas Term 1832.* Oxford, UK: Printed for the Author, 1833.

Longva, P. *A System of Natural Resource Accounts: Rapporter Fra Statistisk Sentralbyra.* Working paper. Oslo, Norway: Statistisk Sentralbyra, 1981.

Lowndes, V., and C. Skelcher. The dynamics of multi-organizational partnerships: An analysis of changing modes of governance. *Public Administration* 76, no. 2 (1998): 313–33. doi:10.1111/1467-9299.00103.

Luck, G. W., G. C. Daily, and P. R. Ehrlich. Population diversity and ecosystem services. *Trends in Ecology and Evolution* 18, no. 7 (2003): 331–36. doi:10.1016/s0169-5347(03)00100-9.

Luck, G. W., R. Harrington, P. A. Harrison, C. Kremen, P. M. Berry, R. Bugter, T. P. Dawson, F. De Bello, S. Díaz, C. K. Feld, J. R. Haslett, D. Hering, A. Kontogianni, S. Lavorel, M. Rounsevell, M. J. Samways, L. Sandin, J. Settele, M. T. Sykes, S. Van Den Hove, M. Vandewalle, and M. Zobel. Quantifying the contribution of organisms to the provision of ecosystem services. *BioScience* 59, no. 3 (2009): 223–35. doi:10.1525/bio.2009.59.3.7.

Madden, J., J. Cohen, T. Baumann, M. Grady, A. Patney, E. Ripley, A. Deitz, S. Gilligan, and S. Feast. Introducing the carbon impact factor: A family of financial instruments to differentiate and reward carbon efficiency in commodity production. *Journal of Environmental Investing* 6, no. 1 (2015): 90–114. http://www.thejei.com/introducing-the-carbon-impact-factor-3/.

Marsh, L. L. Shared governance, collaboration, and innovation. In *Property Rights, Economics, and the Environment*, edited by M. D. Kaplowitz. Stamford, CT: JAI Press, 2000.

Maxwell, D., E. McKenzie, and R. Traldi. *Valuing Natural Capital in Business Towards a Harmonised Protocol.* London, UK: ICAEW, 2014.

McNutt, P. *The Economics of Public Choice.* Cheltenham, UK: E. Elgar, 1996.

McNutt, P. Public goods and club goods. *Encyclopedia of Law and Economics.* 1999. http://encyclo.findlaw.com/0750book.pdf.

MDA. Minnesota Agricultural Water Quality Certification Program. 2015. Accessed January 02, 2016. http://www.mda.state.mn.us/awqcp.

Meuleman, L. *Public Management and the Metagovernance of Hierarchies, Networks and Markets: The Feasibility of Designing and Managing Governance Style Combinations.* Heidelberg, Germany: Physica-Verlag, 2008.

Meuleman, L. *Transgovernance: Advancing Sustainability Governance.* Heidelberg, Germany: Springer, 2013.

Meuleman, L. Owl meets beehive: How impact assessment and governance relate. *Impact Assessment and Project Appraisal* 33, (2014): 1–12. doi:10.1080/14615517. 2014.956436.

Meuleman, L., and I. Niestroy. Common but differentiated governance: A meta-governance approach to make the SDGs work. *Sustainability* 7, no. 9 (2015): 12295–2321. doi:10.3390/su70912295.

Mitchell, M., and M. Newman. Complex systems theory and evolution. In *Encyclopedia of Evolution*, edited by M. D. Pagel. New York: Oxford University Press, 2002.

MN Leg. HF 1231. Minnesota State Legislature. February 9, 2010. Accessed January 03, 2016. https://www.revisor.mn.gov/bills/text.php?number=HF1231&session_year=2009&session_number=0&version=latest.

MN Leg. 2014 Minnesota statutes. Revisor.mn.gov/statutes. Accessed January 03, 2016. https://www.revisor.mn.gov/statutes/?id=17.9891.2014

Moore, J. Business ecosystems and the view from the firm. *Antitrust Bulletin* 51, no. 1 (Spring 2006): 31–75.

Moore, J. Predators and prey: A new ecology of competition. *Harvard Business Review* 71, no. 3 (May/June 1993): 75–86.

Moore, J. *Shared Purpose: A Thousand Business Ecosystems, a Worldwide Connected Community, and the Future.* Concord, MA: First Ecosystem, 2013.

Morrison, E. *Network–Based Engagement for Universities: Leveraging the Power of Open Networks.* Proceedings of 10th PASCAL International Observatory Conference, Brest, France. Purdue Center for Regional Development, Oct. 29–31 2012.

Morrison, E. *"Strategic Doing": A New Discipline for Developing and Implementing Strategy within Loose Regional Networks.* Proceedings of Australia-New Zealand Regional Science Association International Annual Conference, Hervey Bay, Queensland, Australia, 2013.

MPCA EA. Environmental assistance grants awarded. Minnesota Pollution Control Agency. February 03, 2010. Accessed January 03, 2016. http://www.pca.state.mn.us/index.php/about-mpca/assistance/financial-assistance/environmental-assistance-grants-and-loans/environmental-assistance-grants-awarded.html.

Murphy, E., M. M. Murphy, and M. Seitzinger. *Bitcoin: Questions, Answers, and Analysis of Legal Issues.* Washington, DC: Congressional Research Service. Fas.org. October 13, 2015. doi:10.5040/9781472555595.

Müller, A., P. Sukhdev, D. Miller, K. Sharma, and S. Hussain. Towards a Global Study on the Economics of Eco-Agri-Food Systems. TEEBweb.org. May 15, 2015. Accessed January 1, 2016. http://doc.teebweb.org.

Nachira, F., A. Nicolai, P. Dini, M. Le Louarn, and L. R. Leon, eds. *Digital Business Ecosystems.* Luxembourg, European: EUR-OP, 2007.

Natural Capital Coalition. *Natural Capital Protocol Principles and Framework*, June 2015. http://www.naturalcapitalcoalition.org/js/plugins/filemanager/files/NCC_Natural_Capital_Protocol_Principles_and_Framework_brochure.pdf.

Naeem, S., J. C. Ingram, A. Varga, T. Agardy, P. Barten, G. Bennett, E. Bloomgarden, L. L. Bremer, P. Burkill, M. Cattau, C. Ching, M. Colby, D. C. Cook, R. Costanza, F. Declerck, C. Freund, T. Gartner, R. Goldman-Benner, J. Gunderson, D. Jarrett, A. P. Kinzig, A. Kiss, A. Koontz, P. Kumar, J. R. Lasky, M. Masozera, D. Meyers, F. Milano, L. Naughton-Treves, E. Nichols, L. Olander, P. Olmsted, E. Perge, C. Perrings, S. Polasky, J. Potent, C. Prager, F. Quetier, K. Redford, K. Saterson, G. Thoumi, M. T. Vargas, S. Vickerman, W. Weisser, D. Wilkie, and S. Wunder. Get the Science Right When Paying for Nature's Services. Science 347, no. 6227 (2015): 1206-207. doi:10.1126/science.aaa1403.

NCD. The natural capital declaration and roadmap: financial sector leadership on natural capital. Natural Capital Declaration. 2014. Accessed January 01, 2016. http://www.naturalcapitaldeclaration.org/wp-content/uploads/2013/09/NCD-Brochure.pdf.

Neeley, T. Water-quality trading debated former EPA official: 'TMDLs Not for Everyone'. DTN/The Progressive Farmer. September 16, 2015. Accessed January 04, 2016. http://www.dtnprogressivefarmer.com/.

Neilson, B. T., J. S. Horsburgh, D. K. Stevens, M. R. Matassa, and J. N. Brogdon. EPRI's watershed analysis risk management framework (WARMF) Vs. USEPA's better assessment science integrating point and nonpoint sources (BASINS). *American Society of Agricultural and Biological Engineers*, (November 8–12, 2003): 460–70. doi:10.13031/2013.15597. Pp. 460-470 in Total Maximum Daily Load (TMDL) Environmental Regulations–II Proceedings of the 8-12 November 2003 Conference (Albuquerque, New Mexico USA), Publication Date 8 November 2003. (doi:10.13031/2013.15597)

Nelson, H. G., and E. Stolterman. *The Design Way: Intentional Change in an Unpredictable World*. Englewood Cliffs, NJ: Educational Technology Publications, 2003.

Ng, I., and L. Andreu. Special issue: Research perspectives in the management of complex service systems. *European Management Journal* 30, no. 5 (2012): 405–09. doi:10.1016/j.emj.2012.06.003.

Nickerson, J. A., and T. R. Zenger. A knowledge-based theory of the firm—the problem-solving perspective. *Organization Science* 15, no. 6 (2004): 617–32. doi:10.1287/orsc.1040.0093.

Niestroy, I. *Sustaining Sustainability: A Benchmark Study on National Strategies Towards Sustainable Development and the Impact of Councils in Nine EU Member States*. Publication no. 2. Utrecht, Netherlands: EEAC Working Group Sustainable Development, 2005.

Norton, J. Thought experiments. Pitt.edu. November 2015. Accessed January 03, 2016. http://www.pitt.edu/~jdnorton/homepage/research/thought_expt.html.

NRCS. Regional Conservation Partnership Program. 2015. Accessed January 03, 2016. http://www.nrcs.usda.gov/wps/portal/nrcs/main/national/programs/farmbill/rcpp/.

NRCS CIG. Natural Resources Conservation Service. CIG Project Search. 2005. Accessed January 03, 2016. http://www.nrcs.usda.gov/wps/portal/nrcs/cigsearch/national/programs/financial/cig/cigsearch/?svsn=All&projStat e=MN&awardYear=2007%3B2006%3B2005&recState=MN.

OADA. Open Ag Data Alliance. 2014. Accessed January 05, 2016. http://openag.io/.

Oddie, R. Top-down planning from the bottom-up? Vision 2020 and the challenges of participatory environmental governance. *A Dialogue on Development, Displacement and Democracy, Ethics of Development-Induced Displacement.* Working Papers No. 3, 2004. http://www. academia.edu/1136629/Top_Down_Planning_from_the_Bottom_Up_ Vision_2020_and_the_Challenges_of_Participatory_Environmental_ Governance_2004_ 4/29/16

OECD. *Environmental Indicators for Agriculture: The York Workshop.* Paris, France: OECD, 1999.

Olander, L., R. J. Johnston, H. Tallis, J. Kagan, L. Maguire, S. Polasky, D. Urban, J. Boyd, L. Wainger, and M. Palmer. *Best Practices for Integrating Ecosystem Services into Federal Decision Making.* Durham, NC: National Ecosystem Services Partnership, Duke University, 2015. doi:10.13016/M2CH07.

Olson, G. Exactly what is 'Shared Governance'? The Chronicle of Higher Education. July 23, 2009. Accessed January 08, 2016. http://chronicle.com/ article/Exactly-What-Is-Shared/47065/.

Olsson, P., C. Folke, and T. Hahn. Social-ecological transformation for ecosystem management: The development of adaptive co-management of a wetland landscape in southern Sweden. *Ecology and Society,* 2nd ser., 9, no. 4 (2004): 2. http://www.ecologyandsociety.org/vol9/iss4/art2/.

Olsson, P., L. Gunderson, S. Carpenter, P. Ryan, L. Lebel, C. Folke, and C. S. Holling. Shooting the rapids: Navigating transitions to adaptive governance of social-ecological systems. *Ecology and Society,* 18th ser., 11, no. 1 (2006): 18. http:// www.ecologyandsociety.org/vol11/iss1/art18/.

Ostrom, E. "A diagnostic approach for going beyond panaceas. *Proceedings of the National Academy of Sciences* 104, no. 39 (2007): 15181–5187. doi:10.1073/ pnas.0702288104.

Ostrom, E. *Governing the Commons: The Evolution of Institutions for Collective Action.* New York: Cambridge University Press, 2011.

Ott, W. R. *Environmental Indices: Theory and Practice.* Ann Arbor, MI: Ann Arbor Science, 1978.

Patterson, J., K. Schulz, J. Vervoort, C. Adler, M. Hubert, S. Van Der Hel, A. Schmidt, A. Barau, P. Obani, M. Tebboth, K. Anderton, S. Börner, and O. Widerberg. *Transformations Towards Sustainability. Emerging Approaches, Critical Reflections, and a Research Agenda.* Working paper no. 33. Amsterdam, the Netherlands: Earth System Governance Project, 2015.

Phillips, S. T. *This Land, This Nation: Conservation, Rural America, and the New Deal.* Cambridge, NY: Cambridge University Press, 2007.

Plummer, R., D. R. Armitage, and R. C. De Loë. Adaptive comanagement and its relationship to environmental governance. *Ecology and Society E&S* 18, no. 1 (2013): 21. doi:10.5751/es-05383-180121.

Popova, D. Glocalization through internet. *Management and Education* V, no. 3 (2009): 245–248.

Porter, M., and M. Kramer. Creating shared value: How to reinvent capitalism— and unleash a wave of innovation and growth. *Harvard Business Review,* January/February 2011: 3–17.

Porter-O'Grady, T. *Implementing Shared Governance: Creating a Professional Organization.* St. Louis, MO: Mosby Year Book, 1992.

Porter-O'Grady, T. Is shared governance still relevant? *JONA: The Journal of Nursing Administration* 31, no. 10 (2001): 468–73. doi:10.1097/00005110-200110000-00010.

Provan, K. G., and P. Kenis. Modes of network governance: Structure, management, and effectiveness. *Journal of Public Administration Research and Theory* 18, no. 2 (2007): 229–52. doi:10.1093/jopart/mum015.

Rahman, H. M. T., G. M. Hickey, and S. K. Sarker. A framework for evaluating collective action and informal institutional dynamics under a resource management policy of decentralization. *Ecological Economics* 83 (2012): 32–41. doi:10.1016/j.ecolecon.2012.08.018.

Rapacioli, S., L. Lang, J. Osborn, Dr., and S. Gould. Accounting for natural capital: The elephant in the boardroom. May 2014. Accessed January 01, 2016. http://www.cimaglobal.com/Thought-leadership/Research-topics/Sustainability/Accounting-for-natural-capital-the-elephant-in-the-boardroom/.

Redlin, B., M. McLaughlin, W. Fitzgerald, P. Gillitzer, and W. Place. Certifying Minnesota's farms prosper together. Minnesota Agricultural Water Quality Certification Program Report to the Legislature, January 30, 2015. http://www.mda.state.mn.us/news/~/media/Files/news/govrelations/legrpt-mawqc15.pdf.

Repetto, R. C. *Report on Natural Resource Accounting: Information Paper on the Use of Natural Resource Accounting for Countries.* Canberra, Australia: Australian Environment Council, 1988.

Rhodes, R. A. W. The new governance: Governing without government. *Political Studies* 44, no. 4 (1996): 652–67. doi:10.1111/j.1467-9248.1996.tb01747.x.

Rick, T. Organisational culture eats strategy for breakfast and dinner. Torbenrick. eu. June 11, 2014. Accessed January 02, 2016. http://www.torbenrick.eu/blog/culture/organisational-culture-eats-strategy-for-breakfast-lunch-and-dinner/.

Rittel, H. W. J., and M. M. Webber. *Dilemmas in a General Theory of Planning.* Berkeley, CA: Institute of Urban and Regional Development, University of California, 1973.

Roe, M. Governance. In *Maritime Governance and Policy-making*, edited by M. Roe, 41–110. London, UK: Springer-Verlag, 2013.

Roldan, M. D. G. Z. Globalization and glocalization: the Philippine experience. In *Globalization, Governance, and the Philippine State*, edited by J. L. V. Avila, 295–334. Makati City, Philippines: Philippine APEC Study Center Network, 2011.

Rosell, S. *Report of the Roundtable on Renewing Governance Changing Frames: Leadership and Governance in the Information Age.* Report. Ottawa, ON, Canada: Gilmore Printing, 2000.

Ruhl, J. B., S. E. Kraft, and C. L. Lant. *The Law and Policy of Ecosystem Services.* Washington, DC: Island Press, 2007.

Sampson, R. N. *For Love of the Land: A History of the National Association of Conservation Districts.* League City, TX: Association, 1985.

Sampson, Neil, Melinda Kimble, and Sara Scherr. Solutions from the Land. Report. Edited by Dan Dooley and Kent Schescke. Washington DC: United Nations Foundation, 2013. Accessed April 07, 2016. http://www.sfldialogue.net/pathways_report.html.

Sanchirico, J., and J. Siikamäki. Putting a value on nature's services. *Resources*, no. 165 (Spring 2007): 8–11. doi:10.1787/467844512702.

Sandhu, H., S. Wratten, R. Costanza, J. Pretty, J. R. Porter, and J. Reganold. Significance and value of non-traded ecosystem services on farmland. *PeerJ* 3 (2015): e762. doi:10.7717/peerj.762.

Sandhu, H. and Wratten, S. (2013) Ecosystem Services in Farmland and Cities, In Ecosystem Services in Agricultural and Urban Landscapes, edited by S. Wratten, H. Sandhu, R. Cullen and R. Costanza), A John Wiley & Sons, Oxford. doi: 10.1002/9781118506271.ch1.

Sayer, J., T. Sunderland, J. Ghazoul, J.-L. Pfund, D. Sheil, E. Meijaard, M. Venter, A. K. Boedhihartono, M. Day, C. Garcia, C. Van Oosten, and L. E. Buck. Ten principles for a landscape approach to reconciling agriculture, conservation, and other competing land uses. *Proceedings of the National Academy of Sciences* 110, no. 21 (2013): 8349–356. doi:10.1073/pnas.1210595110.

Scharmer, O., and K. Käufer. *Leading from the Emerging Future: From Ego-system to Eco-system Economies.* San Francisco, CA: Berrett-Koehler, 2013.

Schellnhuber, H. J., P. J. Crutzen, W. C. Clark, and J. Hunt. Earth system analysis for sustainability. *Environment: Science and Policy for Sustainable Development* 47, no. 8 (2005): 10–25. doi:10.3200/envt.47.8.10-25.

Schnepf, M., and C. A. Cox. *Managing Agricultural Landscapes for Environmental Quality: Strengthening the Science Base.* Ankeny, IA: Soil and Water Conservation Society, 2007.

Scholz, J. T., and B. Stiftel. The challenges of adaptive governance. In *Adaptive Governance and Water Conflict: New Institutions for Collaborative Planning,* edited by J. T. Scholz and B. Stiftel. Washington, DC: Resources for the Future, 2005a.

Scholz, J. T., and B. Stiftel. Introduction. In *Adaptive Governance and Water Conflict: New Institutions for Collaborative Planning,* edited by J. T. Scholz and B. Stiftel. Washington, DC: Resources for the Future, 2005b.

Selman, M., E. Branosky, and C. Jones. *Water Quality Trading Programs: An International Overview.* Issue brief. Washington, DC: World Resources Institute, 2009.

Shell. *The Shell Global Scenarios to 2025.* Publication. London: Shell International Limited, 2005.

Shields, R. Data is the new currency for publishers. Exchange Wire (blog), April 27, 2015. https://www.exchangewire.com/blog/2015/04/27/data-is-the-new-currency-for-publishers/.

Simon, H. A. *The Sciences of the Artificial.* 3rd ed. Cambridge, MA: MIT Press, 1996.

Smith, A. *An Inquiry into the Nature and Causes of the Wealth of Nations.* Chicago, IL: Encyclopedia Britannica, 1952.

Smith, R. *Users and Uses of Environmental Accounts: A Review of Select Developed Countries.* Washington, DC: World Bank Group, 2014. https://www.waves-partnership.org/.

Snyder, S. *The Simple, the Complicated, and the Complex: Educational Reform through the Lens of Complexity Theory.* OECD Education. Working Papers, no. 96, 2013. doi:10.1787/5k3txnpt1lnr-en. Accessed April 29, 2016. https://www.oecd.org/edu/ceri/WP_The%20Simple,%20Complicated,%20and%20the%20Complex.pdf

SRC. Insight Research: Adaptive Governance. Publication no. 3. January 22, 2015. http://www.stockholmresilience.org/21/news--events/research-insights/insights/2-28-2012-insight-3-adaptive-governance.html.

Sørensen, E., and J. Torfing. Making governance networks effective and democratic through metagovernance. *Public Administration* 87, no. 2 (2009): 234–58. doi:10.1111/j.1467-9299.2009.01753.x.

Sørensen, Eva, and Jacob Torfing. Introduction Collaborative Innovation in the Public Sector & Jacob Torfing. *The Innovation Journal: The Public Sector Innovation Journal* 17, no. 1 (2012): 1–14.

Stafford, S. L. *Private Policing of Environmental Performance: Does it Further Public Goals? Faculty Publications.* Paper 1646, 2012. doi:10.2139/ssrn.1721022. Accessed April 29, 2016. http://lawdigitalcommons.bc.edu/ealr/vol39/iss1/3/

State of MN. Governor Dayton, USDA Secretary Vilsack and EPA Administrator Jackson team up on groundbreaking water quality certification program. News release, January 7, 2012. http://mn.gov/governor/newsroom/pressreleasedetail.jsp?id=102-34735.

Stephens, P. A., N. Pettorelli, J. Barlow, M. J. Whittingham, and M. W. Cadotte. Management by proxy? The use of indices in applied ecology. *Journal of Applied Ecology* 52, no. 1 (2015): 1–6. doi:10.1111/1365-2664.12383.

Stine, J. *Water Panel Discussion.* Proceedings of Agri-Growth Council Annual Meeting, Minneapolis Convention Center. Minneapolis, MN, November 12, 2015.

Stock, P., and R. J. F. Burton. Defining terms for integrated (multi-inter-trans-disciplinary) sustainability research. *Sustainability* 3, no. 12 (2011): 1090–113. doi:10.3390/su3081090.

Stokols, D. Toward a science of transdisciplinary action research. *American Journal of Community Psychology* 38, no. 1–2 (2006): 63–77. doi:10.1007/s10464-006-9060-5.

Stoneham, G. *Creating Markets for Environmental Goods and Services: A Mechanism Design Approach.* Braddon, Australian Capital Territory: Land and Water Australia, 2009.

Stowe, W. *Sixty-day Notice of Intent to Sue.* Lawsuit Filing. Des Moines: Des Moines Water Works, 2014.

Stringham, T. K., W. C. Krueger, and P. L. Shaver. State and transition modeling: An ecological process approach. *Journal of Range Management* 56, no. 2 (2003): 106. doi:10.2307/4003893.

Swihart, D. *Shared Governance: A Practical Approach to Reshaping Professional Nursing Practice.* Marblehead, MA: HCPro, 2006.

Taleb, N. N. *The Black Swan: The Impact of the Highly Improbable.* New York: Random House, 2007.

Tappeiner, G., U. Tappeiner, and J. Walde. Integrating disciplinary research into an interdisciplinary framework: A case study in sustainability research. *Environmental Modeling and Assessment* 12, no. 4 (2007): 253–56. doi:10.1007/s10666-006-9067-1.

Termeer, C., A. Dewulf, and M. Van Lieshout. Disentangling scale approaches in governance research: Comparing monocentric, multilevel, and adaptive governance. *Ecology and Society* 15, no. 4 (2010): 29. URL: http://www.ecologyandsociety.org/vol15/iss4/art29.

Tether. *Tether: Fiat Currencies on the Bitcoin Blockchain.* Working paper. April 2015. https://tether.to/wp-content/uploads/2015/04/Tether-White-Paper.pdf. Accessed April 29, 2016. https://tether.to/wp-content/uploads/2015/04/Tether-White-Paper.pdf

Thomas, L. The Processes of Ecosystem Emergence. Publication. July 2, 2014. http://questromworld.bu.edu/platformstrategy/files/2014/07/platform2014_submission_26.pdf.

Thomas, L., and E. Autio. *Emergent Equifinality: An Empirical Analysis of Ecosystem Creation.* Proceedings of 35th DRUID Celebration Conference 2013. Barcelona, Spain, February 26, 2013.

Thompson, G. *Markets, Hierarchies, and Networks: The Coordination of Social Life.* London, UK: Sage Publications, 1991.

Timmer, C. P. The agriculture transformation. In *Handbook of Development Economics,* edited by H. Chenery, T. Srinivasan, J. Behrman, D. Rodrik, M. Rosenzweig, T. Schultz, and J. Strauss, 276–328. Vol. 1. Amsterdam, The Netherlands: Elsevier Science Publishers B.V, 1988.

Toot, M. SUSTAIN. ECICoop.com. September 2014. http://ecicoop.com/wp-content/uploads/2014/10/Dashboard-SUSTAIN-Memo-10-14.pdf.

TSC. The Sustainability Consortium. 2015a. Accessed January 03, 2016. http://www.sustainabilityconsortium.org/.

TSC. The Sustainability Consortium: What we do. The Sustainability Consortium. 2015b. Accessed January 02, 2016. http://www.sustainabilityconsortium.org/what-we-do/#sthash.SvmOfEIN.dpuf.

UNEP. Environmental governance. In *UNEP Year Book 2011: Emerging Issues in Our Global Environment,* Tessa Goverse and Susanne Bech. Nairobi, Kenya: United Nations Environment Programme, 2011.

UNSD. United Nations Statistics Division—National Accounts. 2015. Accessed January 03, 2016. http://unstats.un.org/unsd/snaama/Introduction.asp.

USDA. *USDA's Use of Conservation Program Technical Service Providers: Hearing,* 109th Cong., 1 (2006) (testimony of Crapo, Mike).

USDA. *Fact Sheet: USDA's Building Blocks for Climate Smart Agriculture and Forestry Fact Sheet,* 2014a. Washington, DC: USDA Office of Communications.

USDA. News release. USDA Reminds Farmers of 2014 Farm Bill Conservation Compliance Changes. July 30, 2014b. Accessed January 2, 2016. http://www.usda.gov/wps/portal/usda/usdahome?contentidonly=true&contentid=2014%2F07%2F0155.xml.

USDA. Office of Communications. USDA Partners with EPA, Offers New Resources to Support Water Quality Trading. News Release No. 0260.15. September 17, 2015. USDA.gov. http://www.usda.gov/wps/portal/usda/usdahome?contentid=2015/09/0260.xml&contentidonly=true.

USDA NRCS. 80 Years Helping People Help the Land: A Brief History of NRCS. Nrcs.usda.gov. 2015. Accessed January 03, 2016. http://www.nrcs.usda.gov/wps/portal/nrcs/detail/national/about/history/?cid=nrcs143_021392.

USDA TSP. What is a technical service provider? Technical Service Providers. 2015. Accessed January 07, 2016. http://www.nrcs.usda.gov/wps/portal/nrcs/main/national/programs/technical/tsp/.

Uuemaa, E., M. Antrop, J. Roosaare, R. Marja, and Ü. Mander. Landscape metrics and indices: An overview of their use in landscape research. *Living Reviews in Landscape Research* 3 (2009): 1–28. doi:10.12942/lrlr-2009-1.

Van Zeijl-Rozema, A., R. Cörvers, R. Kemp, and P. Martens. Governance for sustainable development: a framework. *Sustainable Development* 16, no. 6 (November/December 2008): 410–21.

Vandenbergh, M. P. The emergence of private environmental governance. *Cornell Law Review,* 105, (2013): 130–89. doi:10.2139/ssrn.2411688.

Vandenbergh, M. P. The emergence of private environmental governance. *Environmental Law Institute* (February 2014a), 4410125–0134. doi:10.2139/ssrn.2411688.

Vandenbergh, M. The implications of private environmental governance. *Cornell Law Review Online* 99 (February 2014b): 117–39. http://cornelllawreview.org/files/2014/02/99CLRO117-February.pdf.

Vandenbergh, M. *Vanderbilt University Law School Public Law and Legal Theory: The Private Life of Public Law.* Working paper no. 05-16. 2005. Accessed April 29, 2016. https://www.hks.harvard.edu/m-rcbg/papers/seminars/vandenberg _fall_05.pdf

Vermeulen, S., and B. Campbell. How can sustainability science achieve impact for wicked problems such as climate change? Presentation from the independent science and partnership council of the CGIAR S science forum. Cgspace. cgiar.org. October 17–19, 2011. Accessed January 01, 2016. http://www.slideshare.net/cgiarclimate/vermeulen-campbell-wickedproblemspresentation.

Vincent, J. F. V., O. A. Bogatyreva, N. R. Bogatyrev, A. Bowyer, and A.-K. Pahl. Biomimetics: Its practice and theory. *Journal of the Royal Society Interface* 3, no. 9 (2006): 471–82. doi:10.1098/rsif.2006.0127.

Vries, J. De. Is new public management really dead? *OECD Journal on Budgeting* 10, no. 1 (2010): 1–5. doi:10.1787/budget-10-5km8xx3mp60n.

Waldron, J. Property and ownership. *Stanford.edu.* September 6, 2004. http://plato. stanford.edu/archives/spr2012/entries/property/.

Wellman, B. Little boxes, glocalization, and networked individualism. In *Digital Cities II: Computational and Sociological Approaches,* edited by M. Tanabe, P. van den Besselaar and T. Ishida, 10–25. Berlin, Germany: Springer, 2002.

White House Press Secretary. *Executive Order 13508– Chesapeake Bay Protection and Restoration.* The White House. May 12, 2009. Accessed January 02, 2016. https://www.whitehouse.gov/the_press_office/ Executive-Order-Chesapeake-Bay-Protection-and-Restoration.

Williams, B., and S. Van 't Hof. *Wicked Solutions: A Systems Approach to Complex Problems.* Bob Williams, 2014 New Zealand.

Williamson, O. E., S. G. Winter, and R. H. Coase. *The Nature of the Firm: Origins, Evolution, and Development.* New York: Oxford University Press, 1991.

Wood, S., K. L. Sebastian, S. J. Scherr, and N. H. Batjes. *Pilot Analysis of Global Ecosystems: Agroecosystems.* Washington, DC: World Resources Institute, 2000.

Wratten, S. D., H. Sandhu, R. Cullen, and R. Costanza. Ecosystem services in farmland and cities. In *Ecosystem Services in Agricultural and Urban Landscapes,* S. D. Wratten, H. Sandhu, R. Cullen and R. Costanza, 1–13. New York: John Wiley & Sons, 2013.

Wratten, S. D., and H. Sandhu. Ecosystem Services in Farmland and Cities. In *Ecosystem Services in Farmland and Cities,* edited by Robert Costanza, 3-15. 1st ed. John Wiley & Sons, 2013.

Wright, J. *Natural Resource Accounting—An Overview from a New Zealand Perspective.* Working paper no. 22. September 1990. Accessed January 03, 2016. http://hdl.handle.net/10182/1312. https://www.researchgate. net/publication/40223246_Natural_resource_accounting_an_overview_ from_a_New_Zealand_perspective_with_special_reference_to_the_ Norwegian_experience 4/29/16

Yaziji, M., and J. P. Doh. *NGOs and Corporations: Conflict and Collaboration.* Cambridge, NY: Cambridge University Press, 2009.

Young, O., A. Underdal, N. Kanie, S. Andresen, S. Bernstein, F. Biermann, J. Gupta, P. Haas, M. Iguchi, M. Kok, M. Levy, M. Nilsson, L. Pintér, and C. Stevens. *Earth System Challenges and a Multi-layered Approach for the Sustainable Development Goals.* Issue brief. POST2015/UNU-IAS Policy Brief No. 1. Tokyo, Japan: United Nations University Institute for the Advanced Study of Sustainability, 2014.

Yu, J., Y. Li, and C. Zhao. Analysis on structure and complexity characteristics of electronic business ecosystem. *Procedia Engineering* 15 (2011): 1400–404. doi:10.1016/j.proeng.2011.08.259.

Zagt, R., and N. Pasiecznik. Unraveling the landscape approach. The Landscapes for People, Food and Nature Initiative. September 29, 2014. Accessed January 02, 2016. http://peoplefoodandnature.org/blog/unravelling-the-landscape-approach-are-we-on-the-right-track/.

Index